QGIS By Example

Leverage the power of QGIS in real-world applications
to become a powerful user in cartography and GIS
analysis

Alexander Bruy

Daria Svidzinska

[PACKT] open source
community experience distilled
PUBLISHING

BIRMINGHAM - MUMBAI

QGIS By Example

First published: June 2015

Production reference: 1240615

Published by Packt Publishing Ltd.
Livery Place
35 Livery Street
Birmingham B3 2PB, UK.

ISBN 978-1-78217-467-7

www.packtpub.com

Credits

Authors
Alexander Bruy

Daria Svidzinska

Reviewers
Nyall Dawson

Werner Macho

Commissioning Editor
Julian Ursell

Acquisition Editor
Tushar Gupta

Content Development Editor
Nikhil Potdukhe

Technical Editors
Manan Patel

Rupali R. Shrawane

Copy Editor
Vikrant Phadke

Project Coordinator
Vijay Kushlani

Proofreader
Safis Editing

Indexer
Rekha Nair

Graphics
Abhinash Sahu

Production Coordinator
Melwyn Dsa

Cover Work
Melwyn Dsa

About the Authors

Alexander Bruy is a GFOSS advocate and an open source software developer working on the QGIS project. He also maintains a collection of his personal open source projects. He has been working with QGIS since 2006. Now, he is an OSGeo Charter member and QGIS core developer. Currently, Alexander is a freelance GIS developer and works for companies worldwide.

Daria Svidzinska is an associate professor at the physical geography and geoecology department of the Taras Shevchenko National University of Kyiv, Ukraine. From there, she earned a PhD in geography in 2007. Since then, she has been using FOSS GIS for her research in landscape ecology and spatial planning studies. For the past 5 years, she has been actively involved in academic and professional FOSS GIS teaching and training. In 2014, together with her colleague Alexander Bruy, Daria joined the Geo for All initiative and established an ICA-OSGeo-ISPRS research and education lab at her department (`http://lab.osgeo.org.ua/`). Its main objective is to promote and enhance education, research, and services in the field of open geospatial science and applications.

I would like to thank my coauthor, Alexander Bruy, for his patience, support, and encouragement. I also thank our reviewers, Werner Macho and Nyall Dawson, for their valuable feedback, and the whole editorial team for making the writing process smooth and efficient.

Special thanks goes to my dear mother for patiently putting up with my long working hours, giving me the inspiration and courage to rise above obstacles and constraints.

About the Reviewers

Nyall Dawson is a core developer of QGIS, responsible for thousands of contributions spanning the print composer, symbology, and labeling improvements. He has a master's degree in geospatial information, and is passionate about quality cartography and effective visual presentation of information.

He runs a freelance development consultancy, which focuses on targeted development of QGIS's features. He posts regular updates about his QGIS development work on his blog, which you can find at http://nyalldawson.net.

Werner Macho is an open source software developer working on the QGIS project. After graduating from the University of Natural Resources and Life Sciences in Vienna, Austria, he worked some time there as a scientist. Due to his work on QGIS, he got an opportunity to work for Kartoza in South Africa, but then he went back to his roots and is currently working in the area of water management in Austria. Werner has been working on QGIS since 2007, when he searched for an open source GIS tool to aid his work in floodwater protection plans.

Still, his private computer remains a restricted zone for proprietary software, as he strongly believes in the power of open source and the GPL.

www.PacktPub.com

Support files, eBooks, discount offers, and more

For support files and downloads related to your book, please visit www.PacktPub.com.

Did you know that Packt offers eBook versions of every book published, with PDF and ePub files available? You can upgrade to the eBook version at www.PacktPub.com and as a print book customer, you are entitled to a discount on the eBook copy. Get in touch with us at service@packtpub.com for more details.

At www.PacktPub.com, you can also read a collection of free technical articles, sign up for a range of free newsletters and receive exclusive discounts and offers on Packt books and eBooks.

https://www2.packtpub.com/books/subscription/packtlib

Do you need instant solutions to your IT questions? PacktLib is Packt's online digital book library. Here, you can search, access, and read Packt's entire library of books.

Why subscribe?

- Fully searchable across every book published by Packt
- Copy and paste, print, and bookmark content
- On demand and accessible via a web browser

Free access for Packt account holders

If you have an account with Packt at www.PacktPub.com, you can use this to access PacktLib today and view 9 entirely free books. Simply use your login credentials for immediate access.

Table of Contents

Preface

Welcome to *QGIS By Example*. This book will help you understand the capabilities of QGIS, show you how to work with spatial data and perform the most common analyses, and bring your productivity to a new level with the Processing framework.

QGIS is a very popular and user-friendly open source desktop GIS. It provides many useful capabilities and features and their number is continuously growing. It supports a wide range of raster and vector formats, as well as databases and OGC services. It also integrates seamlessly with other FOSSGIS applications. More and more users all over the world choose QGIS as their primary GIS software.

The book will introduce QGIS 2.8.x, and show you how to properly prepare your data for comfortable work, design beautiful maps, and share them with others. It will also show you how to perform different types of analysis and interpret their results. In the final chapters, you will learn how to become more productive by automating repetitive tasks with the QGIS Processing framework, and how to extend the QGIS functionality by developing a Python plugin.

What this book covers

Chapter 1, Handling Your Data, covers the installation of QGIS, introduces its user interface, and shows you how to customize it. In this chapter, you also get to know how to load data from different sources and assemble it in a spatial database.

Chapter 2, Visualizing and Styling the Data, covers the styling of vector and raster data for displaying. Starting from the basic styling options, we go to the more advanced topics, including rule-based rendering and labeling.

Chapter 3, Presenting Data on a Print Map, shows the QGIS functionality that helps us design beautiful and easy-to-read print maps. You get to know about the print composer and learn how to use its capabilities to create great maps and map books (also known as atlases).

Chapter 4, Publishing the Map Online, explains the preparation of the QGIS project for publishing on the cloud service. This includes creating an account in the QGIS Cloud service, adjusting the project settings, uploading data, and publishing the project.

Chapter 5, Answering Questions with Density Analysis, covers techniques useful when working with large and dense point datasets. In this chapter, you learn how to create raster heat maps from point data, and how to use them to detect the hottest regions and examine spatial distribution patterns. You also get to know another technique called binning, which is an alternative approach to heat maps.

Chapter 6, Answering Questions with Visibility Analysis, demonstrates the techniques and tools used for visibility analysis. You get to know which data is necessary for visibility analysis and how to prepare it to get precise and meaningful results. You also learn how to compute viewsheds and present final results in 3D format.

Chapter 7, Answering Questions with Suitability Analysis, covers approaches and techniques used in suitability analysis. We start by interpreting spatial relationships between different objects, and you learn how to express them in the GIS language. Then you get to know how to perform suitability analysis using raster data. Finally, you learn how to interpret the results of the analysis.

Chapter 8, Automating Analysis with Processing Models, teaches you the functionality of the Processing Graphical Modeler. We start with a general overview of the Graphical Modeler. Then we go through the process of model creation, from the very beginning to the final result—a ready-to-use model.

Chapter 9, Automating Analysis with Processing Scripts, covers scripting with the QGIS Processing framework. We start from basic topics, such as using existing Processing algorithms from the QGIS Python Console. Then we see how to create our own Processing scripts with the required functionality.

Chapter 10, Developing a Python Plugin – Select by Radius, contains the topics required to develop your own Python plugin for QGIS. We start with the basic plugin template, then extend it, and finally explain how to design a plugin GUI with Qt Designer. Also, you get to know how to prepare your plugin for publishing.

What you need for this book
You will need the following software to follow all the examples in this book:

- QGIS 2.8.0 or any later version
- Python 2.7.6 (usually installed with QGIS under Windows or the version available in your system under Unix-like OS)

- Qt and PyQt developer tools
- All of this software can be freely downloaded and works under Linux, Mac OS X, and Windows

Who this book is for

If you are either a beginner or an intermediate GIS user, this book is for you. It is ideal for practitioners, data analysts, and application developers who have very little or no familiarity with geospatial data and software. It will open the doors to the wonderful world of QGIS and teach you how to solve the most common GIS tasks with QGIS.

Conventions

In this book, you will find a number of text styles that distinguish between different kinds of information. Here are some examples of these styles and an explanation of their meaning.

Code words in text, database table names, folder names, filenames, file extensions, pathnames, dummy URLs, user input, and Twitter handles are shown as follows: "We can include other contexts through the use of the include directive."

A block of code is set as follows:

```
( 1) SOURCES = __init__.py \
( 2)            selectradius_plugin.py \
( 3)            gui/selectradiusdialog.py \
( 4)            gui/aboutdialog.py \
( 5)
```

Any command-line input or output is written as follows:

```
Calculator-------------------------------------------
>modelertools:calculator

Raster layer bounds----------------------------------
>modelertools:rasterlayerbounds

Vector layer bounds----------------------------------
>modelertools:vectorlayerbounds
```

New terms and **important words** are shown in bold. Words that you see on the screen, for example, in menus or dialog boxes, appear in the text like this: "Clicking on the **Next** button moves you to the next screen."

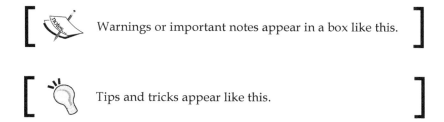

> Warnings or important notes appear in a box like this.

> Tips and tricks appear like this.

Reader feedback

Feedback from our readers is always welcome. Let us know what you think about this book—what you liked or disliked. Reader feedback is important for us as it helps us develop titles that you will really get the most out of.

To send us general feedback, simply e-mail feedback@packtpub.com, and mention the book's title in the subject of your message.

If there is a topic that you have expertise in and you are interested in either writing or contributing to a book, see our author guide at www.packtpub.com/authors.

Customer support

Now that you are the proud owner of a Packt book, we have a number of things to help you to get the most from your purchase.

Downloading the example code

You can download the example code files from your account at http://www.packtpub.com for all the Packt Publishing books you have purchased. If you purchased this book elsewhere, you can visit http://www.packtpub.com/support and register to have the files e-mailed directly to you.

Downloading the color images of this book

We also provide you with a PDF file that has color images of the screenshots/diagrams used in this book. The color images will help you better understand the changes in the output. You can download this file from https://www.packtpub.com/sites/default/files/downloads/4677OS_ColoredImages.pdf.

Errata

Although we have taken every care to ensure the accuracy of our content, mistakes do happen. If you find a mistake in one of our books—maybe a mistake in the text or the code—we would be grateful if you could report this to us. By doing so, you can save other readers from frustration and help us improve subsequent versions of this book. If you find any errata, please report them by visiting http://www.packtpub. com/submit-errata, selecting your book, clicking on the **Errata Submission Form** link, and entering the details of your errata. Once your errata are verified, your submission will be accepted and the errata will be uploaded to our website or added to any list of existing errata under the Errata section of that title.

To view the previously submitted errata, go to https://www.packtpub.com/books/ content/support and enter the name of the book in the search field. The required information will appear under the **Errata** section.

Piracy

Piracy of copyrighted material on the Internet is an ongoing problem across all media. At Packt, we take the protection of our copyright and licenses very seriously. If you come across any illegal copies of our works in any form on the Internet, please provide us with the location address or website name immediately so that we can pursue a remedy.

Please contact us at copyright@packtpub.com with a link to the suspected pirated material.

We appreciate your help in protecting our authors and our ability to bring you valuable content.

Questions

If you have a problem with any aspect of this book, you can contact us at questions@packtpub.com, and we will do our best to address the problem.

1
Handling Your Data

In this introductory chapter, we will prepare all the necessary tools and data for the visual exploration of spatial phenomena, mapping, and analysis. After getting to know some basics about QGIS as a desktop GIS, you will learn how to install and configure it. Then, we will go through the most common spatial data sources and, in the end, aggregate them into a single spatial database that will be used for styling, mapping, and analysis in the following chapters.

In this chapter, we will go through the following topics:

- QGIS installation
- Graphical user interface (GUI), its elements, and its customization
- Loading various types of spatial data from common sources
- Projection's basics
- Assembling data into a single spatial database

Installing QGIS

QGIS is a free and open source desktop GIS on which development began in 2002. Since 2007, the project has been developing under the umbrella of the **Open Source Geospatial Foundation** (**OSGeo**). A relatively young project, QGIS is gaining more and more popularity among individual users, private companies, and organizations all over the world due to the following reasons:

- Distribution under the GNU **General Public License** (**GPL**), which guarantees users the freedom to use, study, share, and modify the software
- Cross-platform support, which means that QGIS can run on Linux, Unix, Mac OS, Windows, and Android operating systems
- Multiple vector and raster data format support, as well as database formats and functionalities

- Permanent improvement of the core functionality, which encompasses data creation, editing, manipulation, analysis, storage, and visual representation
- Permanent growth of the external functionality available from the so-called plugins supported by the international community of developers

 You can find more information about QGIS from its official website at `http://qgis.org/` and its documentation chapter available at `http://documentation.qgis.org/`.

According to the release schedule of the QGIS project, a new version of QGIS is available every 4 months. Starting from QGIS 2.8, every third release has been a long-term release (LTR). It is maintained for a year, and then the next LTR occurs. Throughout this book, an LTR of QGIS 2.8 called "Wein" is used.

Installers for the current QGIS version are available for different operating systems from the download page of the official website at `http://download.qgis.org`. The 32-bit and 64-bit installation files for MS Windows are distributed as follows:

- **QGIS Standalone Installers**: QGIS installation using these has no differences from the conventional software installation procedure under Windows
- **OSGeo4W Network Installer**: Compared to standalone installers, this provides the following advantages:
  - Aside from QGIS, it allows us to install a large number of other packages to work with spatial data (command-line utilities, libraries, and desktop and server applications)
  - It ensures installation of the most up-to-date software versions being directly linked to constantly updating repositories
  - Once defined, the OSGeo4W working environment is convenient to use for automatic and timely updates, package addition, or package deletion

Let's go through the installation procedure of QGIS and its components with the OSGeo4W installer in the example of the latest 32-bit version. Since the installation (especially primary) requires downloading a large number of files, it is desirable to have a high-speed Internet connection. The installation procedure is as follows:

1. Download the latest version of the OSGeo4W 32-bit Network Installer, and double-click on `osgeo4w-setup-x86.exe` to run the installation.

2. In the installer window, select **Advanced Install**, as shown in the following screenshot. From now on, just read carefully and follow the instructions.

3. Choose **Install from the Internet** with the option to keep the downloaded files for future reuse.

4. Choose the directory for the OSGeo4W software suite installation. `C:\OSGeo4W` is a usual and convenient choice. If you want to use another directory, give it a proper name with a short path, excluding spaces and special symbols.

5. Then, choose a directory to save the downloaded packages. You can create a new directory, `C:\OSGeo4W\Downloads`, by typing in its name manually. Keep in mind that the volume of files to be downloaded is about 200 MB and you will need enough space to keep them all.

6. Configure your Internet connection parameters, if any, and click on the **Next** button to proceed. After hitting the **Next** button, the list of all packages is downloaded and you enter the mode of package selection, as shown in the following screenshot. The available packages are grouped into the following categories: **Commandline_Utilities**, **Desktop**, **Libs**, and **Web**.

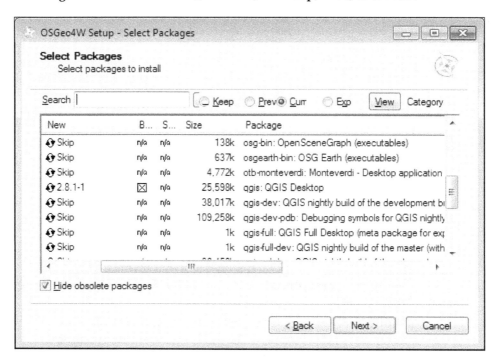

7. To install a package, click on the **+** sign to expand the category, and choose an application by clicking on the rounded arrow beside it (the word **Skip** will be replaced by the latest available version number).

8. To install the latest QGIS version, choose **qgis: QGIS Desktop**, and **grass: GRASS GIS stable release**. GRASS is a standalone GIS application that is closely integrated into QGIS and provides advanced geoprocessing functions.

9. After you've clicked on the **Next** button, the installer automatically detects the necessary dependencies and proceeds to download the packages and installation, as shown in the following screenshot:

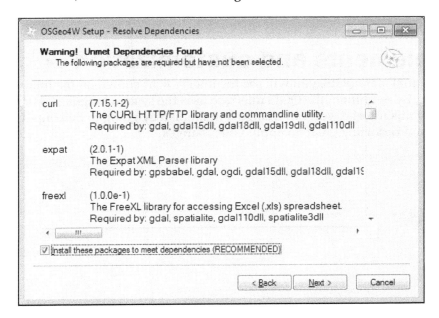

 You can also install **qgis-dev: QGIS nightly build of the master**. Being under constant development, this version is considered unstable and is updated nightly. It is not recommended for workaday use, but is useful if you want to have a try at some new features and give feedback to developers.

After the installation completes, you will find the **OSGeo4W** group in the **Start** menu, with the following components:

- **MSYS**: This is a set of GNU tools that ensure the creation of applications traditionally dependent on UNIX tools

- **OSGe4W Shell**: This is a command-line interface for OSGeo4W utilities

- **QGIS Browser 2.8.1**: This is a QGIS explorer used for navigation and data preview

- **QGIS Desktop 2.8.1**: This is a QGIS desktop application

- **Setup**: This is a OSGeo4W Installer launcher

- **GRASS GIS 7.0.0**: This is GRASS GIS and its components

If you want to update QGIS or expand the current installation with some new components, simply click on the **Setup** shortcut from the **OSGeo4W** menu folder and run the **Advanced installation** procedure. The installer will check for available updates and necessary dependencies, download them, and install them.

GUI elements and customization

When you first start QGIS, you will see the interface, as shown on the following screenshot. By definition, the QGIS interface uses the system language. For the purpose of this tutorial, you can change it in English by going to **Settings | Options | Locale** and restarting QGIS to apply your changes.

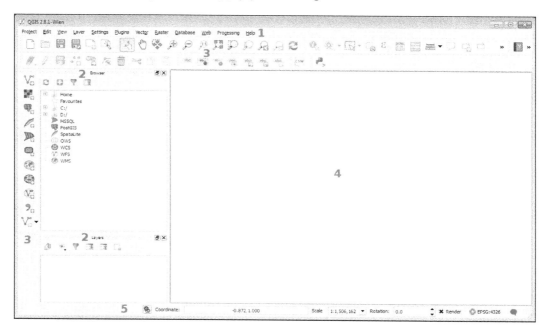

The QGIS interface consists of five main sections:

1. **Menu bar**: This provides access to all QGIS functions in the conventional form — drop-down menus.

2. **Panels**: These are windows that can be either docked or floating. By default, you can see two panels activated on the left side of the workspace.

    ◦   The **Layers** panel is designated to display a tree view list of all loaded layers.

    ◦   The **Browser** panel provides quick navigation and access to the various local, server, or online data sources. You can close or shrink a panel using special buttons in its top-right corner. You can also drag and drop it to a place handily.

Throughout this book, we will use the **Browser** panel from time to time, so try to assign it a proper position instead of closing it. For example, by dragging and dropping it onto the **Layers** panel, you can create two independent tabs: one for layers and the other for data source navigation. In this case, you will not have to sacrifice window the space necessary for expanding and viewing items.

Panels can be turned on or off by going to **View | Panels**. Similarly, you can right-click anywhere on the upper bar of the workspace, and a window with toggles will pop up. This window is divided by a horizontal line and the panels toggles are located in its upper part.

3. **Toolbars**: Menu functions are grouped into logical toolsets and placed as buttons in bars to provide handy access to all the necessary tools. By default, the following toolbars are turned on and displayed:

File toolbar

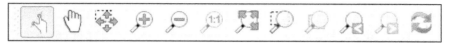

Map Navigation toolbar

Attributes toolbar

Digitizing toolbar

Label toolbar

Help and Plugins toolbars

Manage Layers toolbar

If you want to see a short tool reference, just hover your mouse pointer over the button and a yellow information window will pop up. Some buttons or toolbars are grayed out (for example, the **Digitizing** toolbar and **Label** toolbar), which means that they can be used only after some actions have been performed. For example, the **Digitizing** toolbar's buttons become available only after editing mode is activated for some layers. Toolbars are dockable and can easily be moved around the workspace. To move a toolbar, place the mouse arrow over its border marked by perforation dots. The arrow will turn into a cross, symbolizing that the toolbar can now be dragged and dropped to any other place while holding down the left button of the mouse. You can turn the toolbars on or off by navigating to **View | Toolbars,** or from the window called by a right click on any toolbar.

4. **Map area**: This is the largest section of the interface window, and is designed for data map display and visual exploration.

5. **Status bar**: This briefly displays the following information about your current map overview: a progress bar of the rendering (visible only if rendering takes some time to show its progress), the mouse pointer's position coordinates, the scale, and the current coordinate reference system. It also contains toggles for switching from point position coordinates to extent, and temporarily turning off map rendering. There are also buttons used to open the coordinate reference system dialog [EPSG:4326] and show the **Log Message** panel .

Advanced GUI customization is possible by navigating to **Settings | Customization**. From this dialog, you can get complete control over all menus, panels, bars, and widgets. To start the process of customization, turn on the **Enable customization** toggle. As long as you are not familiar with QGIS interface, it will be easier to enter **Switch to catching widgets in main application** mode. In this mode, you can click on any GUI item you want to hide—a button on a toolbar or a toolbar itself. The chosen items are highlighted and the relevant customization tree item is expanded, as shown in the following screenshot. If you have chosen an item by mistake, simply click on it once again to unselect. Otherwise, you can expand the relevant item, check/uncheck some toggles, and click on **OK**. The changes are applied after QGIS restarts.

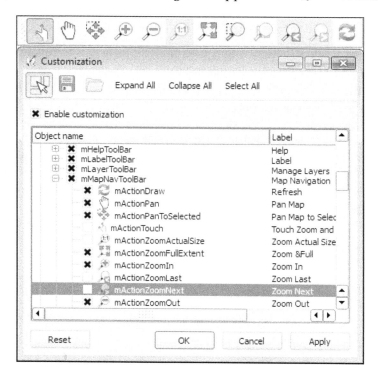

 As your familiarity with QGIS grows and you start using it for different tasks, it might be convenient to create a few users' GUI profiles, aimed at different use cases (for example, digitizing, working with databases or raster data, and so on). To do so, you need to adjust all the necessary settings in the **Customization** dialog window, and save them in a `.ini` file. After that, you will be able to quickly modify the interface just by loading settings from a predefined customization file.

Some minor interface changes, such as general style, default icon size, font and more, are available from the **Application** group, which can be found by going to **Settings | Options | General**.

Extending functionality through plugins

From the very beginning, QGIS has had a modular architecture that makes it easy to add new features or functions. Most functions in QGIS are implemented in so-called plugins, which are divided into the following types:

- **Core plugins**: These are included in QGIS by default and are maintained by the development team. In order to be used, they should only be activated by a user.

- **External plugins**: These are located in an external repository and are maintained by the authors. In order to be used, they should first be installed by a user. With time passing, some of the most useful and popular plugins are incorporated into the QGIS core functionality.

Managing plugins

Plugin management implies their activation, installation, update, and removal, which are performed by navigating to **Plugins | Manage and Install Plugins**. There are several tabs in the dialog window. Clicking on an individual plugin in any of these tabs shows you detailed information: whether the plugin is experimental or not, the functionality, rating, author, version, and also some links to its **homepage**, **code repository**, and **tracker**:

- **All**: Under this tab, you can see the full list of available plugins, both installed and not installed.

- **Installed**: This tab shows the plugins that are already installed in QGIS. If you want to activate/deactivate a plugin, just check/uncheck the toggle beside it.

- **Not installed**: This tab contains a list of all available plugins that are not installed in your QGIS.

  The following tabs are not permanent and are shown only if there are plugins that satisfy some conditions:

  - **Upgradeable**: This tab is visible only when more recent versions of the installed plugins are available in the repositories.

- Invalid: This tab is shown when there are installed plugins that are broken or incompatible with your QGIS version. If you click on an individual plugin from this tab, you will be shown the information about the possible cause of invalidity; so that you would be able to provide a consistent feedback to the developers.

- New: This tab shows the not-installed plugins seen for the first time.

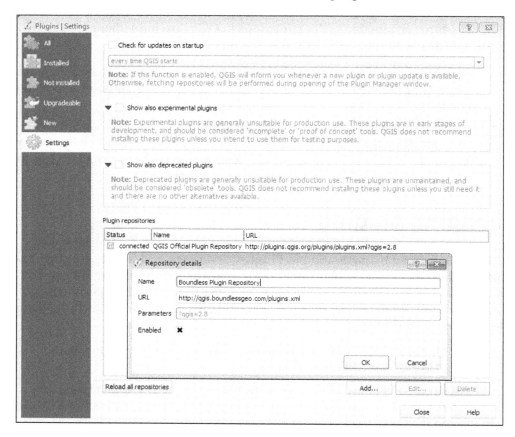

- **Settings**: This tab allows you to define how often updates will be checked and whether to use experimental or deprecated plugins. While it is not recommended to use experimental plugins for production purposes, you can still activate the option to explore the entire range of available tools. By default, only **QGIS Official Plugin Repository** is connected, but you can connect additional repositories if you know them by clicking on the **Add...** button. For example, the previous screenshot shows how to add an external Boundless plugin repository.

To install a plugin, go back to the **Not installed** tab, select the plugin you want to install, and click on **Install plugin**. After that, the installation starts, and upon its completion, the message about successful installation is displayed. The installed plugin is activated by default and either appears in the **Plugins** menu, or is added to another relevant menu (for example, **Vector** or **Raster**). Also, some plugins appear as separate toolbars and panels that can be placed and turned on or off manually.

Finding and choosing a plugin that fits your needs might be difficult because their number is over 400 and is constantly growing. To explore the world of QGIS plugins, you can start from `http://plugins.qgis.org/`, which contains information about popular and featured plugins. If you are looking for a specific function but don't know the exact plugin name, we advise you to search by using tag words in the **Manage and Install Plugins** dialog window, as shown on the following screenshot:

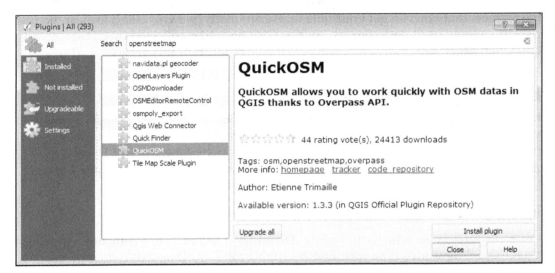

 By default, plugins are sorted alphabetically by name. If you want to rate them (by downloads, votes, or status), apply the additional sorting options available by right-clicking contextual menu of the list of plugins.

Loading data into QGIS

The data in the GIS can be submitted as separate files, databases, or external online sources. Moreover, there are different data models used in the GIS to represent the geometry of spatial objects. The vector data model is mostly used to express discrete features, such as points (for example, separately growing trees and points of interest), lines (roads or railways), and polygons (buildings or administrative borders). Raster data is convenient for expressing continuous phenomena that are better represented by coverage than by points, lines, or polygons. The most common raster data sources are remote sensing imagery, digital elevation models, and scanned and georeferenced topographic maps.

QGIS uses the GDAL/OGR spatial data library to read and write multiple vector and raster file formats. In the following sections, we will briefly discuss the most common file formats that you can get your data in. Additionally, we will take a look at the recently widely used data sources, such as CSV files, files collected by GPS-receivers, and OpenStreetMap.

Loading shapefiles

Most of our data is in the ESRI shapefile format, which is one of the most common vector data file formats. There are a few alternative ways of loading a shapefile into QGIS. You can open the dialog window by going to **Layer** | **Add Layer** | **Add Vector Layer**, or by clicking on the relevant button $V_{\square}$ on the **Manage Layers** toolbar, or by using the *Ctrl + Shift + V* keyboard shortcut.

In the **Add vector layer** dialog window, configure the following items:

- **Source type**, which can be **File**, **Directory**, **Database**, or **Protocol**. By default, the **File** source type specified is suitable for loading spatial data represented by separate files, like in our case.

- Consider the **Encoding** drop-down list options if the data you work with contains special symbols or the charset of its textual attributes differs from the conventional Latin symbology. In this case, you should choose an appropriate encoding type from the drop-down list. Our data does not contain any special symbols, so we accept the default **System** encoding.

- The **Browse** button is meant for navigating to the work directory. There, you can choose one or several files (with the *Ctrl* key) to add after hitting the **Open** button.

For the first time, you are likely to be overwhelmed by the amount and diversity of the available files. This is because in the GIS, a dataset is usually represented by several files with different extensions. For example, an ESRI shapefile consists of at least 4 files (.shp, .shx, .dbf, and .prj) that share the same name, and among them, .shp is what you should select to load the data into QGIS. You see all the files because the file type filter in the bottom-right corner of the window is set to **All files** by default. To hide all unnecessary files, choose the **ESRI shapefiles** file type. In future, don't forget to adjust the filter according to the desired file type.

Alternatively, you can use the **Browser** panel to navigate to the working directory. Select the files you want to load (either one or several, by holding down the *Ctrl* key), and then just drag and drop them onto the map area. If you want to simplify the navigation, add the folder you are currently working with to **Favourites** from its right-click menu shortcut.

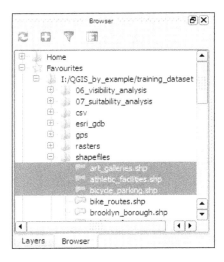

The loaded files are assigned random colors. Don't worry about that as of now; we will cover layer symbology in *Chapter 2, Visualizing and Styling the Data*.

Loading rasters

The procedure of adding raster files is similar to what was described earlier, so we will briefly go through it for the example of the common raster file format **GeoTIFF**:

1. You can load a raster file by going to **Layer** | **Add Layer** | **Add Raster Layer**, by clicking on the relevant button ▓ on the **Manage Layers** toolbar, or by using the *Ctrl + Shift + R* keyboard shortcut.

2. In the **Open a GDAL Supported Raster Data Source** window, go to the work directory.

3. To see available the GeoTIFF files, remember to adjust a file type filter, select one or more files (by holding down the *Ctrl* key), and click on **Open**. Alternatively, you can drag and drop rasters from the **Browser** panel.

Loading data from the Personal GeoDatabase

ESRI Personal GeoDatabase is an original ArcGIS data format where all of the database's content is held in a single .mdb Microsoft Access file. It is widely used to store data in a single data file instead of multiple files of different formats.

Adding an ESRI file geodatabase to QGIS is similar to adding any other vector format:

1. Access the dialogue window by going to **Layer** | **Add Layer** | **Add Vector Layer**, or by clicking on the relevant button V on the **Manage Layers** toolbar, or with the *Ctrl + Shift + V* keyboard shortcut.

2. Go to the work directory, set the file type filter to **ESRI Personal GeoDatabase**, and select the database file you want to import data from.

3. In the **Select vector layers to add...** window, you will see the available layers and their characteristics. Click on the layers to be added and hit **OK**.

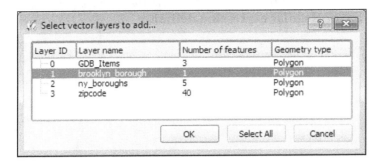

 Raster layers contained in the database will be interpreted and loaded into QGIS as bounding-box polygons. If the data is stored in tables that do not contain any spatial features, the geometry type will be displayed as **Unknown**. The tables will be loaded and available for exporting to other formats (for example, .dbf) or joining with spatial data layers.

Importing CSV files

A **Comma-separated values (CSV)** file is another popular data file format. In fact, it is just a spreadsheet with field values delimited by commas. There are various delimiters possible instead of commas (for example, tabs, spaces, colons and so on). Very often, these tables contain spatial data in the form of positional attributes represented by point longitude or latitude (XY) coordinates, or **well-known text (WKT)** geometry that describes points, lines, or polygons.

In our dataset, we have a CSV file called noise.csv. It contains the details of noise complaints registered mostly by the New York City Police Department. To add this file as a spatial layer, follow these steps:

1. Open the **Create a Layer from a Delimited Text File** dialog window by going to **Layer | Add Layer | Add Delimited Text Layer**, or just click on the relevant button ⁹ₒ in the **Manage Layers** toolbar.

2. After browsing and pointing to your file, QGIS tries to parse it using the specified delimiter. By default, the comma delimiter is used, but you can specify any other delimiter using **Custom delimiters** (comma, tab, space, and so on) or **Regular expression delimiter**.

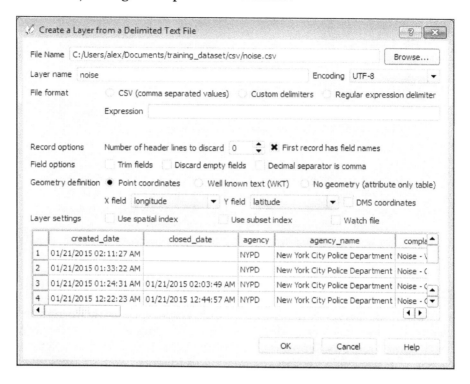

3. The dialog also provides access to a number of other useful settings; for example, turning on **First record has field names** creates headers for fields. After defining the geometry as **Point coordinates**, **X field** and **Y field** containing **Longitude** and **Latitude** values will be loaded from the dataset automatically.

4. After you have clicked on **OK**, QGIS reads the data and might show a delimited text file error message similar to **Errors in file full_path_to_the_file/filename.csv 100 records discarded due to missing geometry definitions**. This means that some points do not contain geographic coordinates, and so they cannot be located properly. After closing the message window, you will be prompted to specify the **Coordinate Reference System (CRS)** for the layer.

As we can see from the **Latitude** and **Longitude** column values, point coordinates were originally registered by the receiving devices of the **Global Position System** (**GPS**) in decimal degrees. It is also known that GPS uses the WGS 84 CRS. This is why in the **Coordinate Reference System Selector** window, we enter the **EPSG: 4326** code filter and specify **WGS 84** under **Geographic Coordinate Systems** as the initial CRS. After clicking on **OK**, you will see that the data is rendered in the map canvas as points.

QGIS uses CRS definitions based on the European Petroleum Search Group's (EPSG) Geodetic Parameter Dataset, which contains detailed structured descriptions of coordinate reference systems and transformations of global, regional, national, and local applications. The database of EPSG identifiers can be used to specify a CRS in QGIS. You can read more about the EPSG Geodetic Parameter Dataset at http://www.epsg-registry.org/.

Loading GPS data

As GPS receivers became portable and relatively cheap, GPS tracking became a ubiquitous and widely used technique of collecting data during a field survey, or simply tracking your own routes while running or cycling. Depending on the receiver's capabilities, in addition to spatial coordinates, a lot of information can be registered and written—for example, time, elevation, and so on. Registered information is stored in points that represent location changes (way, track, or route points), a planned route (if it exists), and a track of movement. There are numerous formats for storing GPS data. As primary data format, QGIS uses the standard interchange **GPX (GPS eXchange)** format.

We will load the `nyc_marathon.gpx` file from our dataset. You can load a `.gpx` file as a vector layer by going to **Layer | Add Layer | Add Vector Layer**, by clicking on the relevant button V in the **Manage Layers** toolbar, or by using the *Ctrl + Shift + V* keyboard shortcut. Go to the data directory and apply the **GPS eXchange format (GPX)** file type filter. Select the file (or files) you want to add and click on **Open**. Alternatively, you can drag and drop files from the **Browser** panel.

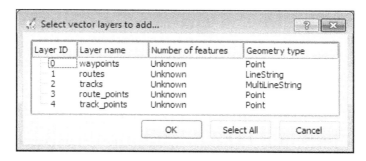

As GPS data consists of points, routes, and tracks, the window pops up where you choose exactly which feature you want to open. You can select one or more features by clicking on a row and holding down the *Ctrl* key. If you are not sure exactly which feature you need, use **Select All**. Every feature type will be presented in a separate layer. By turning layers on and off, you can check whether they contain data or not; for example, the file we work with contains only tracks and track points. You can remove unnecessary empty layers by right-clicking on the layer contextual shortcut **Remove**.

Advanced options for working with GPS data are available through the **GPS Tools** core plugin. After activation, the plugin functions can be accessed by navigating to **Vector | GPS | GPS Tools**, or from the button on the **Vector toolbar** panel.

Getting OpenStreetMap data

OpenStreetMap (OSM) is a crowdsourcing project aimed at the development of an open map of the world. Worldwide, the OSM community uses high-resolution remote sensing imagery, GPS surveying, and local knowledge to make the map as accurate and detailed as possible. As the OSM data is licensed under Open Data Commons' **Open Database License (ODbL)**, it has recently become a widely used source of spatial data. Understanding the importance of unrestricted access to spatial data, QGIS supports OSM, providing integrated access and support for its data.

The QGIS core functionality is available from the **OpenStreetMap** menu under **Vector**. Here, you can find all the required tools to download the data and export it into conventional spatial data formats, but if you are new to the QGIS and OSM data concepts, you may find the process tedious. This is why we recommend that you use external **QuickOSM** plugin capabilities to download the OSM data.

After installation as described in the *Extending functionality through plugins* section, the plugin's main functionality can be accessed by going to **Web | QuickOSM | QuickOSM**. The **Quick query** tab of the main window contains everything you need to perform a query and get the data into QGIS. First, define a `key=value` pair from the drop-down lists, or just type it in manually.

 It can be quite difficult to select the appropriate tags for the first time if you are not familiar with the OSM tagging system. Each tag is a `key=value` pair that describes some point, linear, or polygonal feature. The key describes a broad class (for example, amenities), while the value gives the details (bank, cinema, café, bicycle parking, and so on). Use the **Help with key/value** button to know more about tagging from the OSM wiki page at `http://wiki.openstreetmap.org/wiki/Mapfeatures`.

Then, set up the area of interest extent by location name, map canvas, or layer. If the layer drop-down list is inactive, click on the button beside it to refresh it. In the **Advanced** section, you can turn on/off different geometry types. By default, all geometry types are turned on, so all the relevant points, lines, and polygons will be downloaded. Browse to the directory you want to save the data in. In the following screenshot, you can see the example of the QuickOSM query used to download bicycle parking point data.

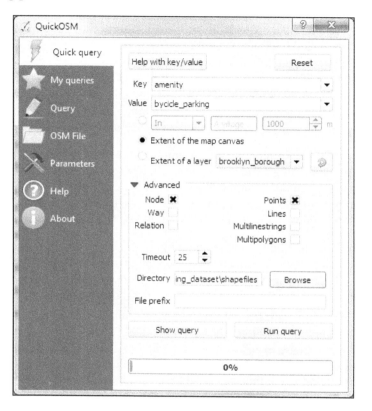

> If you are familiar with the **Overpass API** query language, you can click on **Show query** or the **Query** tab to modify the initial query.

After clicking on the **Run query** button, the shapefile with the data will be loaded into the map canvas.

Dealing with projections

Projections define how real-world objects on the curved surface of the earth will be flattened and projected on a map-like planar surface. Different data sources are usually created and distributed in different projections, depending on acquisition techniques and the scope of application. To be able to manipulate and analyze them properly in QGIS, it is important to understand how it interprets and manages information about projections.

QGIS supports about 2,700 CRS. They constitute a database, each item of which is described by an **ESPG identifier**, and a description line in the format of the **PROJ.4** projection library. To store and read information about projection, QGIS uses its own format stored in .qpj files. There are two important points to keep in mind while working with projections: a data source projection and project projection—which are not always the same.

Data source projection

When you load your data into QGIS, it tries to automatically define information about the projection from the accompanying projection description file and set it to a newly added layer. You can check the layer's projection information in the **Layer Properties** dialog. To open it, double-click on the layer name in **Layers** panel, or use the right-click contextual shortcut **Properties**. Information about the projection is shown in the **Coordinate reference system** section of the **General** tab, like this:

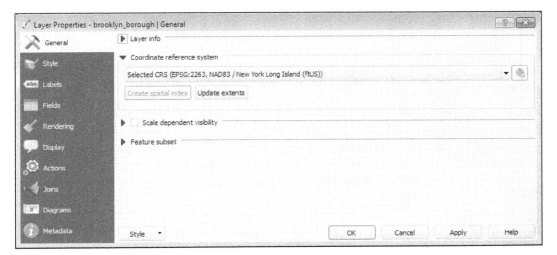

There are numerous projection description file formats, and they mostly depend on the data format and origin. One of the most common among them is a `.prj` file. This file format sometimes provides information about the projection in a reduced form, and QGIS cannot define it properly. If information about the projection is absent or incomplete, you will see that the projection description is missing or looks like this: `USER:100000 - * Generated CRS (+proj=etc.)`. It means that QGIS didn't find an appropriate projection within its own database and added it as a custom CRS. If you are sure that the projection is standard but QGIS fails to interpret it properly because of a reduced description, specify it manually. There are different ways to assess the **Coordinate Reference System Selector** dialog as follows:

- Double-click on the layer name in **Layers** panel to open the **Layer Properties** dialog. Under the **General** tab, click on the **Select CRS** button 🌐 to select the necessary CRS from the **Coordinate reference system selector** dialog window.

- By going to **Layer | Set CRS of Layer(s)**.

- Using the *Ctrl + Shift + C* keyboard shortcut.

- By right-clicking on the layer shortcut **Set Layer CRS**.

In the **Coordinate reference system selector** dialog, choose the necessary CRS from the list and click on **OK**. Note that you can use a filter by key words or the EPSG code to hide unnecessary projections. With time, often used projections will be added to the list of the recently used CRS.

If QGIS doesn't define a projection for your data automatically and you are not sure about projection name and EPSG code, use the Prj2EPSG website at `http://prj2epsg.org/`. In this service, you just need to upload your `.prj` file, and well-known text projection information from it will be converted into standard EPSG code.

Note that specifying the projection manually from the **Properties** dialog is a temporary solution; it does not persist after closing a file and/or exiting QGIS. This means that whenever you are loading this file into QGIS, you will have to specify the projection again. If you are going to work with the file in other projects, it is better to use the **Define Current Projection** tool by going to **Vector | Data Management Tools | Define Current Projection**. In this dialog, you can choose a predefined spatial reference system from the **Coordinate reference system Selector** dialog or **Import spatial reference system from existing layer** with a known CRS. In both the cases, the existing `.prj` file will be overwritten by QGIS with all the necessary parameters, and you will not have to adjust the projection manually next time.

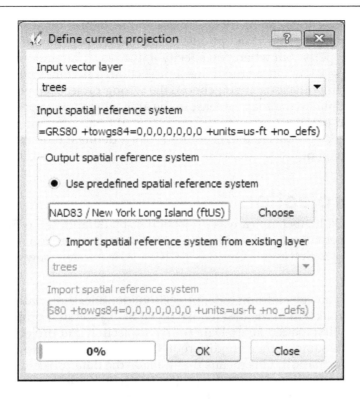

All of this is true when you import files from external sources. If you create a shapefile in QGIS, both `.prj` and `.qpj` files are created, so you don't need to define the projection manually. The files projection will be interpreted correctly in the other GIS applications that read it from `.prj` files.

You can also maintain QGIS behavior when loading or creating a new layer that has no CRS information by going to **Settings | Options | CRS**, where three alternatives of CRS for new layers are available:

1. **Prompt for CRS**: In this case, you will always be asked to manually specify the CRS for the layer.

2. **Use project CRS**: The data will be automatically assigned a CRS set for a project. This is a convenient choice if you are using files in the same projection, but if their projection differs from what is specified, you will not be able to see them.

3. **Use default CRS displayed below**: Conventionally, **EPSG:4326 - WGS 84** is used, but you can specify your own choice.

Working with projections can be tricky, unless you recognize the difference between the **specify** and **define projection** options. It's all about assigning information about the projection properly, but when you specify it, the result is temporary within a current working session. Once you define a projection, it becomes permanent unless redefined again. In both cases, if you choose the wrong projection, you will not be able to see your data on the map (at least in the area where you expect it to be). Usually it is good practice to specify the projection at first, comparing your data to the correctly projected dataset. If everything is okay, define it to make projection assignment permanent.

Project projection

When you load QGIS, its map canvas projection is, by default, set to **EPSG:4326 – WGS 84**. This behavior is defined by the global default projection, which can be changed by going to **Settings | Options | CRS | Always start new projects with this CRS**.

However, when you start adding data, projection identifier automatically changes to the projection of the first loaded layer. So, by default, project projection is defined by the first loaded layer. Data source projection and project projection are not always the same. If they are different, QGIS can apply the so-called **on-the-fly transformation**. This means that it reads the data projection, and if it differs, automatically aligns it with the project projection, applying the necessary transformations. There are a few ways to specify the projection for the project and enable on-the-fly transformation:

- By going to **Project | Project Properties | CRS**.

- Click on the **Current CRS** ⊕ EPSG:4326 button on the **Status** bar, and activating **Enable 'on the fly' CRS transformation** in the **Project properties | CRS** window.

- Go to **Settings | Options | CRS | Default CRS for new projects**. Activate either **Automatically enable 'on the fly' reprojection if layers have different CRS** or **Enable 'on the fly' reprojection by default**.

In the last way, the difference is very subtle and is mainly obvious when adding the first layer to the project. In the case of automatically enabled reprojection, the default map canvas CRS will be superseded by the first layer projection, and the following layers will be automatically adjusted to it too. In the case of reprojection by default, all layers will be reprojected into the project initial CRS (this is usually default global projection, unless you've changed it).

After on-the-fly transformation is enabled, you will see an **OTF** abbreviation near the current CRS on the status bar, and the icon will become solid.

If you want to quickly change the project CRS according to a specific layer in the **Layers** panel, right-click on a layer and choose the **Set Project CRS from Layer** shortcut.

Loading layers to a spatial database

Now that we have loaded all of the available data into QGIS, let's aggregate it into a single database. QGIS supports working with various database management systems and their spatial extensions (PostgreSQL/PostGIS, Oracle Spatial/GeoRaster, MSSQL Spatial, SQL Anywhere, and SQLite/SpatiaLite). Among them, PostGIS and SpatiaLite are the ones you have probably already heard about. Being common, these spatial databases are usually used for different purposes. PostGIS is an example of an enterprise solution used mostly on a server to provide spatial data maintenance and access for multiple users. SpatiaLite is a lightweight file database for personal use. Usage of SpatiaLite database has a number of advantages as follows:

- All of the data is stored in a single, portable file, and you don't get overwhelmed by different file types (for example, `.shp`, `.dbf`, `.shx`, `.prj`, `.qpj`, and so on)
- Shapefile limitations for size (up to 2 GB) and field name length (10 characters) can be omitted
- Built-in spatial indices allow you to perform searches for large areas faster
- This is a real relational database in a single format that allows you to use various spatial functions and SQL-based workflows

Perform the following steps to assemble your data into a single SpatiaLite database:

1. First of all, you need to create an empty spatial database to load your data into. You can do this from the **Browser** panel. Right-click on the **SpatiaLite** entry and choose the **Create database** shortcut. Specify the folder and a `.sqlite` database file name. You will get the **The database has been created** message. Now you can expand the entry and see your newly created empty database.

2. We will use the DB Manager core plugin that provides a single interface to work with different databases to load the data. The plugin functionality is accessible by going to **Database | DB Manger**. On the left side of the window, you can see a tree list of available database connections grouped by type. Once created, your database will automatically be available under the **SpatiaLite** item, and you can connect to it by double-clicking. When the connection is established, you will see general information about the database or its selected items in the **Info** tab on the right side of the window. To import files into the database, click on the **Import layer/file** button 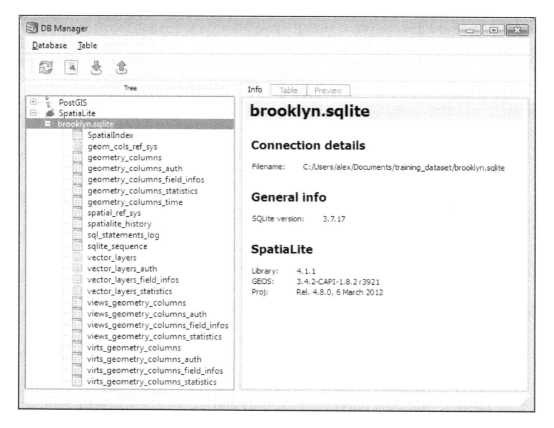, or access it from the **Table** menu.

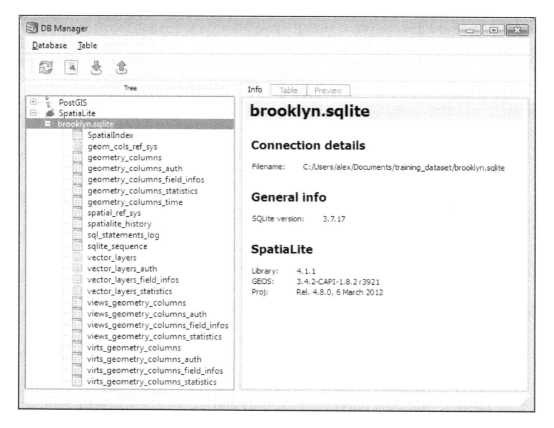

3. In the opened **Import vector layer** window, you should define the layer you want to import. It could be one of the already loaded layers available in the **Input** drop-down list, or the layer you choose from the filesystem by clicking on the browse button **....** Then, define the **Output table** name (in our case, it is ny_boroughs, the same as the input).

The following options allow you to exercise more control over your data:

- ° **Primary key**: If this is not specified, the field will be named pk by default

- ° **Geometry column**: If this is not specified, the field will be named geom by default

- ° **Source SRID** and **Target SRID**: These are CRS EPSG codes read from the data, but you can specify them manually if the data isn't accompanied by projection files, or if you want to reproject it before loading it to the database

- ° **Encoding**: If not specified, the dataset charset is set to **UTF-8** by default

- ° **Drop existing table**: If you import a table with the same name as a previously existing table, it will be replaced by the new table

- ° **Create single-part geometries instead of multipart**: Multipart features will be disaggregated into single-part geometries before loading to the database

- ° **Create spatial index**: A spatial index that allows faster spatial search and query performance will be created

After hitting **OK** and waiting a little (depending on the file size, the process may take some time), you will get the `Import was successful` message. To explore the imported table click on the **Refresh** button , and you will see a geometry table in the list of database items. From the **Info** tab, you can view information about table, geometry type, fields, and triggers. The **Table** tab displays data in a tabular form, and the **Preview** tab displays spatial geometry, if any. By right-clicking on any element, you can delete, rename, or add it to the map canvas. All other opened layers can be loaded in a similar way.

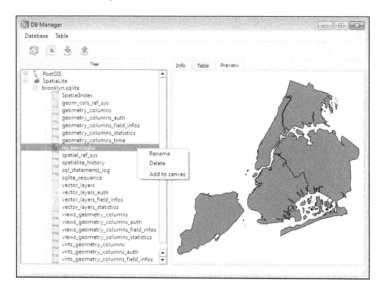

> Aside from the core functionality, there is an external plugin called **QSpatiaLite** that is specially designed to support work with SpatiaLite databases. After installation, it will be available when you navigate to **Database | SpatiaLite | QSpatiaLite**.

Summary

For now, we have covered a few introductory topics and completed several tasks. First of all, you learned how to install QGIS, configure its interface, and extend functionalities with plugins. Then we discussed common data sources you can use to get data into QGIS. You are now capable of exploring and managing data in different projections. Last but not least, you imported all your data into a portable, lightweight SpatiaLite database file.

In the next chapter, we will move forward and reveal QGIS's potential for data visualization.

2
Visualizing and Styling the Data

Collecting and organizing data from different sources is only half of the story. The next step is to present it on a map, accurately disclosing thematic content and features. For this purpose, various techniques of visual design, or styling layers based on their attributes, are used in GIS. QGIS has incredibly broad and flexible capabilities of cartographic visualization and data styling.

In this chapter, we will go through the following topics and skills:

- Good practices with respect to data organization within a single working document (project)
- Visual representation of data, revealing its thematic and spatial features
- Layer labeling, which is done to enhance data readability
- Control over styles
- Base maps used to provide spatial context and a background for your data

The main result of this chapter will be a project, designed according to the basic requirements of cartographic visualization.

Loading layers from the spatial database

Throughout this chapter, we will use the database that we created earlier. As you remember, this is a SpatiaLite file database. In order to work with it, we must first establish a connection and then load the data. As always, there are a few alternative ways of doing this in QGIS:

- From the menu, go to **Layer** | **Add Layer** | **Add SpatiaLite Layer**
- Click on , the **Add SpatiaLite Layer** button, in the **Manage Layers** toolbar
- Use the *Ctrl + Shift + L* keyboard shortcut

In the opened window, define the database you want to connect to by following these steps:

1. If you have created the database as was described in the previous chapter, you are likely to notice that it has already been connected. If it is not, then click on the **New** button and navigate to the .sqlite file you want to work with.

2. After the database file is selected, click on the **Connect** button to see the list of available layers. You will see the layers' names and geometry types.

3. Select the necessary layer (or layers, which is done by holding down the *Ctrl* key) and click on the **Add** button.

> The **Set filter** button gives you more flexibility when adding the data, as it provides access to the **Query Builder** window, where you can make up a conditional expression to define the sub-dataset, as shown in the following figure. For example, if you want to select a multipurpose play area (MPPA) from the athletic facilities dataset, you can do the following:
>
> - On the left side of the **Query Builder** window, select a necessary field called primary_sp (double-click on its name to add to the expression in the **Provider specific filter expression** textbox)
> - Type in or click on the operator button (the equality sign in our case)
> - Load the field values by clicking on the **All** button, and double-click on an MPPA value to add it to the expression
> - In the **Provider specific filter expression** textbox, you will see the following line: "primary_sp" = 'MPPA'
>
> Test the query. If it returns a meaningful result, click on **OK**. After returning to the main window, you will see the condition beside the **geom** column. Click on the **Add** button to load the layers.

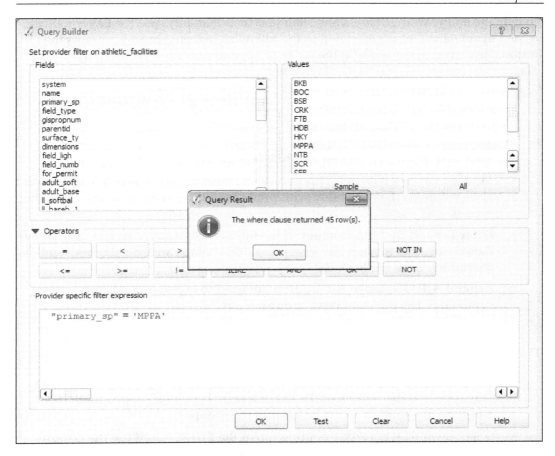

Loading layers is even simpler from the **Browser** panel. To see the available layers, successively expand **SpatiaLite** and the relevant database item. Now you can just drag and drop the layers onto the map canvas.

Grouping and reordering layers

By default, layers are loaded in alphabetical order (reversely if you add them from the browser panel). Each new layer is laid on top of the previous layer and covers it. By default, all layers are turned on, styled with simple uniform symbols, and randomly assigned colors. The order of layers can be changed by simply dragging and dropping them up and down the legend. Also, it is strongly advisable to order and arrange layers in some logical groups, as it simplifies the navigation and understanding of data.

To manage and rearrange layers and maintain their visibility, use the **Layer** toolbar in the **Layers** panel, which is shown in the following screenshot and contains the buttons described afterwards (in the order of appearance from left to right):

- **Add Group**: This creates an empty layers group.
- **Manage Layer Visibility**: This allows us to quickly show and hide layers and also customize their visibility with predefined layer combinations – so-called presets.
- **Filter Legend By Map Content**: If the filter is active, layer legend only shows those items that are actually visible inside the map canvas. All other symbols are hidden from the legend.
- **Expand All / Collapse All**: Buttons are used to expand or collapse layers, symbology legends, and layer groups, if any.
- **Remove Layer/Group**: This removes the selected legend entries.

There are two ways of creating a layer group:

- Click on 📋, the **Add Group** button, from the **Layers** toolbar. A new group will appear at the bottom of the list of layers. Type in an appropriate name and then drag and drop the layers into the group.
- Select several layers while holding down the *Ctrl* key, and use the **Group Selected** right-click contextual shortcut to place them in a single group.

It is possible to develop a multilevel hierarchy with subgroups by selecting groups and applying **Add Group** to them. Any item of the layer legend, whether it's a single layer or a group, can be renamed with the **Rename** right-click shortcut. Renaming doesn't affect the dataset itself, but allows us to apply to it a proper and meaningful name within a project. Now try arranging the layers into several meaningful groups yourself, and name them appropriately.

Developing your own styles

In QGIS, a style is a way of cartographic visualization that takes into account a layer's individual and thematic features. It encompasses basic characteristics of symbology, such as the color and presence of fill, outline parameters, the use of markers, scale-dependent rendering, layer transparency, interactions with other layers, and labels.

A well-chosen style greatly simplifies data perception and readability, so it is important to learn how to work with styles in order to be capable of representing your data in the best way. In this section, we will discuss vector and raster layers separately, as their stylization has some unique features.

Developing styles for vector layers

The **Style** menu from the **Layer Properties** dialog provides you with all the necessary tools to symbolize and style your data. To open it, double-click on a layer name in the **Layers** panel, or use the **Properties** right-click contextual shortcut, and select the **Style** section. You will see something similar to this:

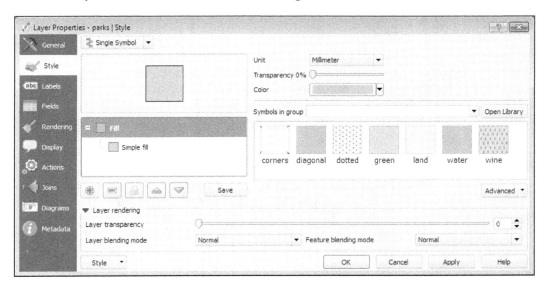

The very first thing you should pay your attention to is the small **Renderer** drop-down type list in the top-left corner. It contains the following items:

- **Single Symbol**: This is the simplest type that draws all the layer features with the same symbol.
- **Categorized**: This defines data-driven categories and allows us to symbolize them individually.
- **Graduated**: This defines the categories based on quantitative attributes, allowing us to rank features gradually.
- **Rule-based**: This is the most flexible and advanced renderer type. It allows the user to define their own categories using multiple criteria and style them individually.

- **Point displacement**: This renderer is useful when you are working with a point layer containing overlapping points that have similar coordinates or are located too close to each other. It is available only for single point layers, and will automatically shift the location of the markers so that all overlapping markers are visible.

- **Inverted polygons**: These are used to style the area exterior to the polygon, and are available for polygon layers only.

- **Heatmap**: This represents a point layer with a continuous surface according to the points' density, It is available for point layers only.

Depending on the renderer type selected, the **Style** menu section changes its view, and if you mistakenly select an inappropriate type (for example, the heat map renderer for a line layer), you will get the corresponding message.

After the renderer type is chosen, you can start adjusting the symbology with the **Symbol selector** dialog, whose accessibility depends on the renderer type selected. For example, for the **Single symbol** and the **Inverted polygons** renderer, this dialog is directly available from the **Style** section. For the **Categorized** and **Graduated** renderers, it is available from the **Symbol Change** button, which looks like this:

Symbol [   ] ▢ Change...

Rule-based, **Point displacement**, and **Heatmap** renderers have their own specifications for symbology selection and adjustment, which will be covered in the following sections. Nevertheless, regardless of the renderer type selected (except the **Heatmap** renderer), you always have access to the **Symbol selector** dialog, which looks similar to what is shown in the following screenshot and consists of several sections:

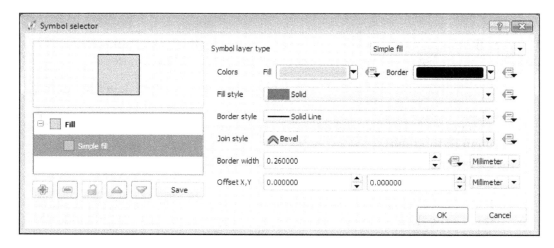

In the top-left part of the window, you can see a symbol preview. Underneath the preview, there are symbol layers. By default, only one layer is used, but you can add more with the **Add symbol layer** button, ⊕, or remove unnecessary layers with the **Remove symbol layer** button, ▭, which is active only if there are two or more layers available. With the **Lock layer's color** button, ⬚, the layer's color will be locked for changes, which prevents colors from being modified by a categorized or graduated renderer. Layers can be reordered with the **Move up** (△) and **Move down** (▽) buttons, and if you are satisfied with the result, **use the Save symbol button** in the symbol library.

In the right-hand side of the dialog window, there are options available for the symbol layer selected. Among them, **Symbol layer type** is the most important. The list of available types depends on the layer geometry. For a polygon layer, you can select from the following list:

- **Centroid fill**: Polygons are symbolized by a marker at the polygon's centroid, instead of rendering the entire area of the polygon. This is useful if you have a lot of small polygons that are better to visualize by a point than by a tiny polygon that is visible only after zooming in.

- **Gradient fill**: Use a predefined gradient or create a custom gradient to fill the polygon.

- **Line pattern fill**: Line patterns can be combined to create various hatching effects. These effects are useful when you want to use the same fill color underneath but want to highlight some differences between objects with hatching.

- **Point pattern fill**: Regularly distributed points (or other symbols) can fill the polygon and create a pattern.

- **Raster image fill**: Any raster image can be used to create a background-filling texture or pattern.

- **SVG fill**: A scalable vector graphics `.svg` file (or a marker) can be used to create a filling texture or pattern.

- **Shapeburst fill**: This shades the interior of a polygon, depending on the distance from the edge of the polygon, and creates amazing border buffering effects.

- **Simple fill**: This the default type, and is characterized by fill color, pattern, and border.

- **Outline: marker line**: A marker symbol is used as the outline.

- **Outline: Simple line**: Only the polygon's outline is drawn and its properties defined by line color, width, and style.

For a point layer, you can choose from different marker types represented by **Ellipse**, letters or signs (**Font marker**), various markers (**Simple marker**), icons (**SVG marker**), or attribute field values (**Vector Field marker**).

For a line layer, the **Simple** and **Marker line** types are available. In the first case, the line is rendered as usual, and in the second, a regularly recurring marker symbol is used. **Marker line** can be used, for example, to show a line's direction (movement on a road, a river flow, and so on) with an arrow marker symbol.

Now that we have covered the basics, we will closely explore different rendering styles in the example of some layers from our database.

Styling a layer with the Single Symbol renderer

In this example, we will use the Brooklyn borough boundaries layer to outline the area of interest on the map. As you can see in the following screenshot, we use two symbol layers, both defined as **Outline: Simple line**, but with different **Pen style** patterns. Underneath, we place a lighter **Solid line** and cover it with a darker **Dash Line** pen style. Selecting contrast colors allows us to implement a border-like stripe effect.

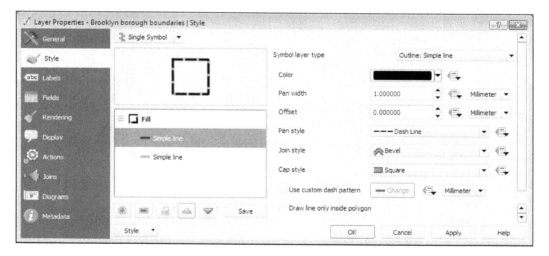

Styling a layer with the Categorized renderer

Let's style the public schools layer to reflect the school type:

1. After setting the renderer type to **Categorized**, select the column with the categories to be rendered under **Column**. The column is available from the drop-down list that contains all the layer attribute fields. As we want school types to be shown as categories, select the sch_type field to categorize the layer:

2. Click on the **Symbol Change** button to adjust the layer symbology. In the **Symbol selector** window, set **Symbol layer type** to **SVG marker** and navigate to the training dataset catalog called svg. Select the school.svg file and adjust the symbol's **Size** value to 3 mm. Click on the **OK** button to come back to the main window:

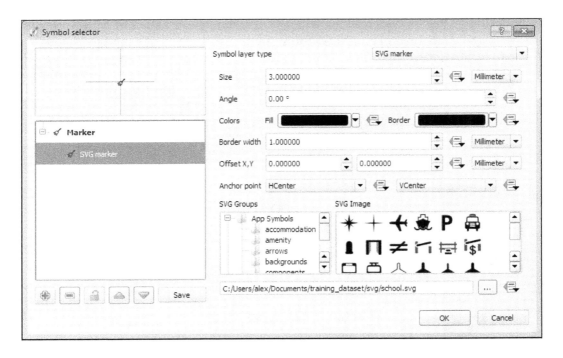

3. Click on the **Classify** button. The window with the **Symbol**, **Value**, and **Legend** columns will automatically be populated with the categories and their descriptions from the attribute field. The difference between **Value** and **Legend** is that values represent unique attributes, while the legend provides their descriptive characteristics. It's more obvious when values, for example, are some codes and the legend is supposed to explain their meaning. Also, values are not visible in the layer's legend, but their descriptions are. You can add or delete the categories, or manually edit the text of its elements (**Value** or **Legend**) by double-clicking on the item in the relevant column. For now, just click on **OK** to exit the **Style** dialog and take a look at the preliminary result.

4. You will see that the symbols on the map canvas are too dense and overlapped, which makes the map cluttered. To enhance readability, we should relate the appearance of symbols with a scale, and there are two basic approaches to do this:

   ° Activate **Scale dependent visibility** from the **General** tab of the **Properties** window, where you should enter values of **Minimum (exclusive)** and **Maximum (inclusive)** scales. For example, if you define the values, as shown in the following screenshot, the layer will become visible within a scale range of 1:49,999 (because the minimum value of 50000 is excluded) and more. This approach doesn't affect the symbol's size. It only regulates the scale range for its appearance while zooming in and out of the map canvas:

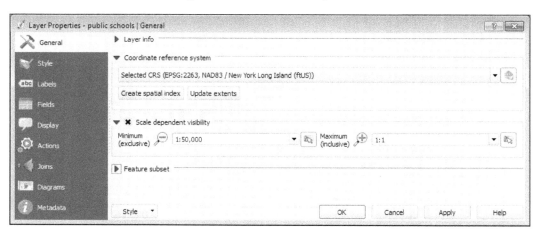

○ The other approach, which we will actually use, is more sophisticated because it makes the symbol size dependent on the scale. To apply it, open the **Style** dialog in the **Properties** window, and click on Symbol [✐ Change...], the **Symbol Change** button, to open the **Symbol selector** dialog, as shown. Activate the necessary symbol layer, change its **Size** units to **Map unit**, and enter 30. This means that the symbol size will be set 30 feet and changed according to the scale while zooming in or out. Click on the **OK** button to come back to the main window.

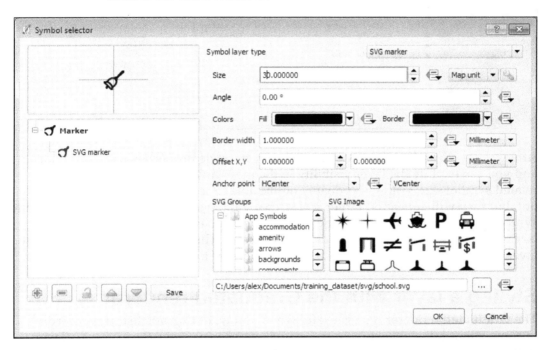

In this case, the scale range can be defined through the **Adjust scaling range** button (), which opens a dialog window to define the minimum and maximum scale values.

5. By default, all categories are assigned the same **Symbol** marker according to the SVG file selected, but you can modify it by double-clicking and adjusting the necessary properties (that is, **Fill** and **Border**) in the **Symbol selector** dialog. Also, right-clicking on any item brings up a contextual menu with the **Copy**, **Paste**, **Change color**, **Change transparency**, **Change output unit**, and **Change size** shortcuts for simplifying some common actions.

This screenshot shows what the screen may look like if you choose the **Categorized** renderer and adjust the settings as described previously:

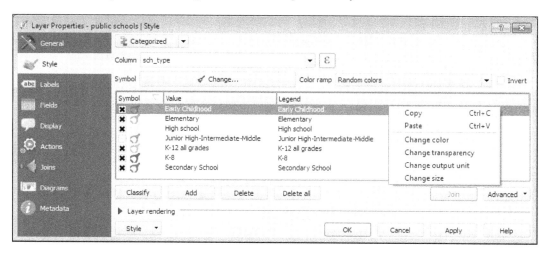

After you've clicked on the **OK** button, the layer's style properties will be applied, and you can explore the layer visibility in the map canvas. In the case of map unit related symbol size, you will probably not be able to see markers while zooming out, but they will appear on larger scales, changing in size accordingly. Note that the symbols in the layer's legend on the **Layers** panel behave similarly, that is, become larger on larger scales and shrink on smaller scales. This helps you know whether the layer is visible on this scale, and if it is, helps you see the appearance of its markers.

Styling a layer with the Graduated renderer

The **Graduated** renderer type is useful when you want to grade features according to some quantitative attribute. In our sample dataset, we have the zipcodes layer, which contains data about population in the population field. We will use this layer and field to rank ZIP code boundaries according to the number of people living and demonstrate how graduated rendering works.

 Instead of choosing an attribute field, you can click on the **Expression dialog** button ε, beside it. It will be useful if you need to use some values that are not in the attribute table, but can be derived from the field values. For example, you can use the area/ 43560 expression to convert area in square feet into acres.

1. Set the renderer type to **Graduated** and select the `population` field in the **Column** section.

2. Define the number of classes you wish to show (normally, five to seven classes are recommended, otherwise it can be difficult to distinguish between them visually), and select a **Color ramp**.

You can use a predefined standard color ramp or select from the advanced options. To use them, select **New color ramp** from the bottom of the drop-down list. In the **Color ramp type** window, you will get access to the following options:

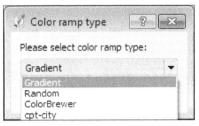

- **Gradient**: This provides options to create and modify custom gradients.
- **Random**: This creates a random color ramp according to various customizable options, such as **Hue**, **Saturation**, **Value**, and **Classes**.
- **ColorBrewer**: Use predefined color palettes for maps designed for optimal clarity. This includes several color schemes for different kinds of data (sequential, divergent, and qualitative).
- **cpt-city**: This provides access to dozens of color gradients for cartography, technical illustration, and design.

3. Then, select an appropriate graduation mode, which is as follows:

   ○ **Equal interval**: The value range is divided into equal range classes according to the number of classes set (for example, values from 0 to 100 are divided into five classes of 20 units each).

   ○ **Quantile (Equal Count)**: All of the data will be divided into the number of classes set, and ranges will be chosen in such a way that each class will contain the same number of items.

   ○ **Natural Breaks (Jenks)**: This method groups values according to their similarity. So the values within a class have minimum variance, but feature values across classes vary significantly.

- ○ **Standard Deviation**: Classes are divided according to the standard deviation of the values, and this shows how the data differs from its mean value.

- ○ **Pretty Breaks**: This creates *n+1* classes for the given range of values (that is, if 5 classes are set, the resulting amount will be 6), similar to **Equal interval**, but breaking points are selected so that the values are nicely rounded numbers (for example, multiples of 10 if using integers).

After you've clicked on the **Classify** button, the **Symbol**, **Values**, and **Legend** columns will be populated. Double-click on any of these to modify it. Right-clicking opens a contextual menu. It is strongly advisable to play around with the various classes and classification modes before deciding which is best for revealing the features of your data.

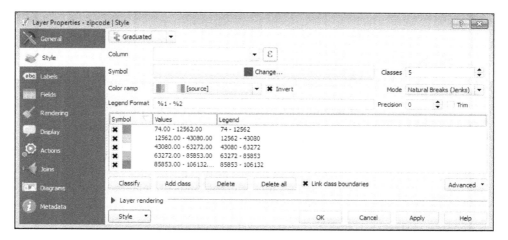

Styling a layer with the Rule-based renderer

The **Rule-based** renderer is the most flexible renderer because it allows you to divide data into your own complex categories and style them separately. We will explore the rule-based rendering capabilities in the example of the `bike routes` layer, which has a lot of attributes that are useful to visualize. For example, there are two fields that are of particular interest to visualize on map. The first one is `allclasses`, which contains the classification of the route types (for existing cycling facilities), based on the NYC Bicycle Master Plan classification. The types are as follows:

- **I**: Greenway/multi-use path.

- **II**: On-street striped bicycle lane.

- **III**: On-street signed bicycle route.

- **0**: Planned, but not yet existing as a bike route.
- **L**: Link. There is no facility present, but the element is suggested as a connection between portions of the bike network.
- **S**: Stairs and pedestrian overpasses.

The second field is `lanecount`, which shows the number of lanes. Rules need to be used to combine these attributes into separate categories that show both the route type and lane count. Follow these steps to develop a Rule-based symbology for the layer:

1. Navigate to **Properties | Style** and select the **Rule-based renderer** type from the drop-down list. To create a rule, click on the **Add rule** ⊕ button. The **Rule properties** dialog window will open.

2. In the window, click on the **...** button beside the **Filter** line. The **Expression string builder** window will open. This window consists of three sections:
   - The left section is **Expression**, with the row of operator buttons above it. You can either type the expression manually or double-click on field names, values, and functions to compose it.
   - The central section of the window is **Functions**. To expand any item, click on the **+** sign beside it and double-click to add it to the expression. We're mostly interested in the **Field** and **Values** items that contain all the available attribute field names. When you highlight any of them and click on the **Load values** buttons that are **all unique** or **10 samples**, the list of values is returned. Again, you can either type a value in the expression manually or add it by double-clicking on it.
   - If you're lost, look at the right part of the window. It contains contextual help for the item highlighted in **Functions**.

In our expression, we want to show greenways and multi-use paths that are in the **I** category. The `"allclasses"` = `'I'` expression for this is shown in the following screenshot:

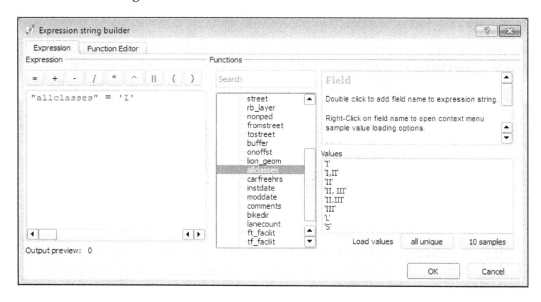

3. After clicking on the **OK** button, you'll be in the **Rule properties** window again. First of all, test the consistency of the expression. You will get the message **Filter returned 'some number' of feature(s)**. If the number of features is 0, it means that there are no features that satisfy the rule. Otherwise, you'll get a positive number that shows the number of features in the class.

4. Then add a **Label** and, if necessary, **Description** to the class. If the **Scale range** toggle is activated, you can define the maximum-minimum scale range to visualize the class on a map. Class symbology is developed in the **Symbol** section of the window as shown:

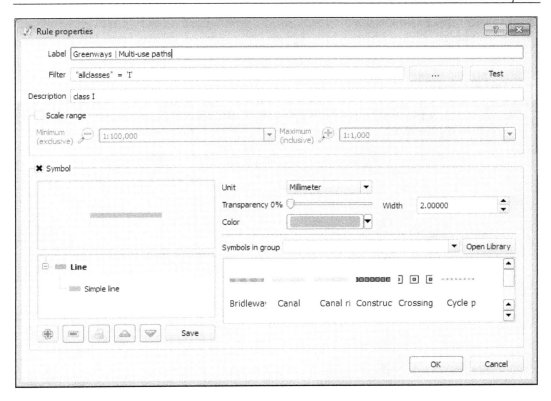

After clicking on the **OK** button, you will be back to the main **Style** window, where rules can be managed with the **Add rule** ⊕, **Edit rule** � , or **Remove rule** ▬ buttons. Go ahead and add categories for the other classes available in this layer.

With the **Refine current rules** button, you can go further and develop multilevel hierarchy based on scales, categories, or ranges. For example, we can take `allclasses` as a major classification attribute and then divide it into subcategories by the number of lanes:

1. Select category, navigate to **Refine current rule | Add categories to rule**, and select the **lanecount** field from the **Column** drop-down list. Click on the **Classify** button:

    Similarly, add subcategories to other rules.

2. Now, we should check our rules to remove empty categories. Select all the available categories by holding down the *Ctrl* key, and click on the **Count features** button. Empty categories will be identified by the **0** count. Click on the **Remove rule** button to exclude them from the legend.

3. As a result, your rule-based style may look like this:

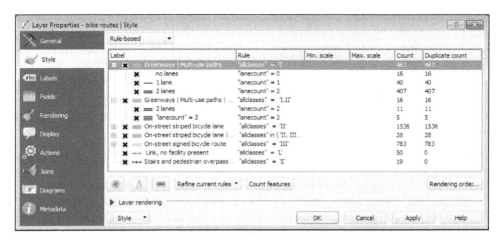

4. You can greatly improve the visualization of multilevel symbols by clicking on the **Rendering order** button and adjusting the symbol levels. In the opened **Symbol Levels** dialog window, you can define the order in which the symbol layers are rendered. The numbers in the cells decide in which rendering pass the layer will be drawn. The idea is that some symbol levels should have a lower rendering order and others should have higher rendering order. Lower levels will be covered by higher ones, allowing the user to create smooth conjunctions and preserve the symbology hierarchy. Set up the rendering order as shown in the following screenshot and explore the results:

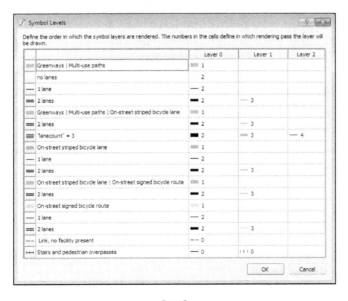

Styling a layer with the Point displacement renderer

Some point layers (for example, the `wifi public` layer from the training dataset) contain points with the same or similar coordinates that are impossible to distinguish on the map because of their overlapping. In this case, **Point displacement** renderer can be of great use, as it locates overlapping points within an imaginary circle. It moves points to a small distance in such a way that they all become visible. Note that the renderer doesn't affect real point positions in the dataset, but visually changes placement to satisfy cartographic requirements. To be able to use this type of rendering, a layer should be a single point feature type.

To apply **Point displacement** renderer to the `wifi public` layer follow these steps:

1. Choose the **Center symbol** marker around which the points will be displaced. By clicking on the symbol button, you get access to the conventional **Symbols selector** dialog. In the window, set **Symbol layer type** to **SVG marker**, navigate to the training dataset catalog called `svg`, and select the `Wi-Fi-Logo.svg` file. Set its size to `20` and select **Map unit** as the size measurements unit. After clicking on the **OK** button, you will be brought back to the main window.

2. Go to **Renderer | Single symbol** for the points themselves. After clicking on the **Rendering settings** button below, you will be in the **Symbol selector** dialog. Again, set the symbol layer type to **SVG marker**, navigate to the training dataset catalog called `svg`, and select the `wi-fi.svg` file. Set its size to `30` and select **Map unit** as the size measurement unit. After clicking on **OK**, you will be taken back to the main window.

3. The **Rendering circles** section defines the properties of the circles that appear on a map to group the points that have similar coordinates and are thus grouped around the common center. The following properties are included:

   ° **Circle pen width**: This defines the outline width of the bordering circle

   ° **Circle color**: This defines the outline color of the bordering circle

   ° **Circle radius modification**: The bigger this value, the bigger the circle will be

   ° **Point distance tolerance**: The bigger this value, the more the points to be snapped around the common center

For now, we accept the default values and apply the settings. In the following screenshot, you can see what the result may look like in the case of several points that share similar coordinates and need to be grouped around a common center:

The **Labels** section contains some major labeling settings, for example, **Label attribute, font, color**, and so on. We temporarily omit this, as labeling will be covered later in the *Adding labels* section of this chapter.

Styling a layer with the Inverted polygons renderer

The **Inverted polygons** renderer styles the polygon outside its boundary and allows us to achieve remarkable cartographic effects. One of its common uses is for showing water bodies with the "fading out" blur effects on the map. For this purpose, **Duplicate** the NY borough boundaries layer with the right-click contextual shortcut, right-click to **Rename** it to water area, open the **Style** properties dialog, and select the **Inverted polygons** renderer.

Now, create a complex symbol that consists of two layers:

1. The first **Symbol layer type** is set to **Outline: simple line**.
2. The second one is **Shapeburst fill**.

   ° In the **Gradient colors** section, we choose the **Two color** gradient from blue to white.

   ° In **Shading style**, activate **Shade to a set distance:**. The bigger the value, the larger the shading effect will be.

○ Finally, adjust the **Blur strength** slider to a value of 10 to make blurring softer. Click on the **OK** button and enjoy the result!

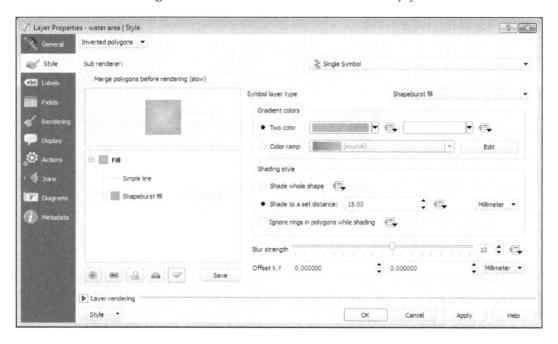

Styling a layer with the Heatmap renderer

Heatmap is a relatively new renderer type. It was introduced in the QGIS 2.8. It represents a point as a continuous-density surface and allows us to apply cool styling effects. Note that in order to be able to apply this type of renderer, the layer and map canvas should be in the same coordinate reference system.

We will apply the renderer to the `trees` layer, which contains over 15000 points. These are hard to show on the map individually because of their high density:

1. After selecting the **Heatmap** renderer type from the drop-down list, select a predefined color ramp for the layer called **Greens**. If you are not satisfied with the color ramp, you can modify it by clicking on the **Edit** button beside it, or create your own color ramp by choosing **New color ramp** from the very end of the list, and finally **Invert** the color ramp if there is a need.

2. The **Radius** field determines the search area for density estimation. It basically represents how close points must be to each other to influence the heat map. The larger the radius, the smoother the surface; and the smaller the radius, the finer the details in the heat map. You can define the radius of a point in different units: **Pixels**, **Millimeter**, or **Map unit**. Remember that map units are scale dependent. Pixels and millimeters change regardless of the scale, but reflect zooming effects as well.

3. **Maximum value** is usually set automatically and is responsible for the maximum density of points per area unit, but you can adjust it to your needs.

4. Also, you can use a **Weight points by** numerical field just in case you want the layer to reflect some important information in addition to density. For example, if we have the relevant data, we can weigh schools by the number of students, or cafés by the number of visitors per month.

5. The **Rendering quality** slider is used to adjust the smoothness of the surface. The higher the quality you choose, the slower the rendering process will be, while coarser surfaces are rendered faster.

Layer rendering

Whichever rendering type you choose, there are always the same **Layer rendering** options at the bottom of the **Style** dialog window. The first one is a general **Layer transparency** percentage slider. You can move it to adjust the layer transparency. **Layer blending mode** provides some special graphics effects for the layer you work with to interact with the bottom layer (or layers). By default, the **Normal** mode is set, which means that the underlying layer is hidden underneath without mixing of colors with covering layer. This is written as (a, b) = a, where a is the top layer and b is the bottom layer.

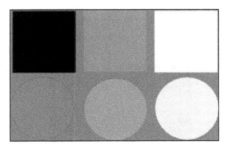

Example of the Normal blend mode applied to the overlapping layers

You can choose from among 12 different blending modes divided into four groups, as described in the following table:

Blend mode	Formula and description	The example of effects
Lighten modes—this group of blending modes makes black disappear, preserves white, and lightens midtones.		
Lighten	*max (a, b)* The maximum of RGB components from both layers are selected.	
Screen	*1-(1-a)×(1-b)* At first, RGB components of both layers are inverted. Then they are multiplied, and the final result is inverted again.	
Dodge	*b÷(1-a)* the bottom layer is divided by the inverted top layer.	
Addition	*a+b* This summarizes components for both layers.	

Blend mode	Formula and description	The example of effects
Darken modes: this group is opposite to the previous one, which means that it preserves black, makes white disappear, and darkens midtones.		
Darken	*min (a, b)* The minimum of RGB components from both layers are selected.	
Multiply	*a × b* The layers are multiplied.	
Burn	*1 - (1-b)÷a* The bottom layer is inverted and divided by the top layer, and the result is inverted again.	

Blend mode	Formula and description	The example of effects
Contrast modes: the modes drop midtones, and apply dark and light pixels from the top layer to darken or lighten pixels from the bottom layer respectively.		
Overlay	Combines two blend modes for different pixels. For lighter pixels, half-strength Screen mode is applied, and for darker pixels, it is Multiply mode. As a result, gray midtones become invisible. Calculations are based on the bottom layer, which means that pixels of the top layer are lightened or darkened by the bottom layer.	
Soft light	This is the same as the previous one, but instead of Screen and Multiply modes, Dodge and Burn modes are used for lightening and darkening respectively.	
Hard light	This is similar to Overlay, but the top and bottom layers are switched.	

Blend mode	Formula and description	The example of effects
Inversion and cancellation modes: apply simple arithmetic actions to invert colors and suppress black		
Difference	*b-a* This subtracts pixel values of the top layer from the bottom layer, preserving only positive values; that is, blending with black (0) has no changes, while blending with white (1) inverts the color.	
Subtract	*b-a* This is similar to the Difference mode, but negative values are substituted by black.	

Feature blending mode applies the same effects to the overlapping features within the layer. We advise you to give the different modes a try in order to understand how they work with different data types and styling options. Getting familiar with them allows you to create professional-looking maps with stunning cartographic design effects.

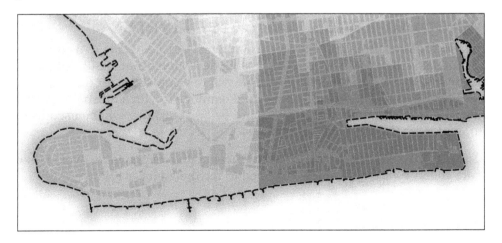

The preceding screenshot is an example of the hurricane evacuation zones layer combined with the underlying primary residential zoning layer through simple 50 percent transparency (on the left), versus the **Multiply blending mode** used without any transparency (on the right).

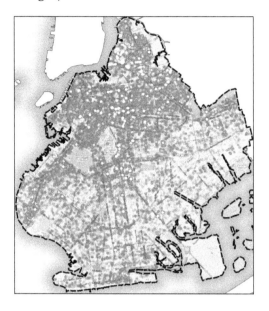

In the preceding screenshot, the `noise` point layer with simple marker fill opacity set to 20 percent; the **Layer blending mode** is **Hard light** and **Feature blending mode** is **Dodge**.

Developing styles for raster layers

Our dataset contains three raster layers: `landcover 2010`, `height a.s.l., ft`, and `hillshade`. All of these are remote sensing derivatives that are useful for representing land use and relief features. We will develop styles for them to represent the DEM as a continuous layer, and combine it with a hillshade to obtain some semi-3D cartographic effects. The `landcover 2010` layer will be used to show how to work with the set of discrete classes.

First, let's style the `landcover 2010` layer that contains several landcover classes. These are coded by integer values described in the metadata as follows:

- 1: `tree canopy`
- 2: `grass/shrub`
- 3: `bare earth`

- 4: `water`
- 5: `buildings`
- 6: `roads`
- 7: `other paved surfaces`

Let's develop the styles by performing the following steps:

1. Open the **Style** tab from the **Properties** dialog window (double-click on the layer in the legend, or select the **Properties** right-click contextual shortcut).

2. Set **Singleband pseudocolor** as **Render type**. As our raster has only one band, it will be loaded automatically as **Band 1 (Gray)** in the **Band** drop-down list.

3. As we work with discrete landcover classes, we go to **Color interpolation | Discrete**.

4. Now we need to add our classes into the legend window. Use the **Add values manually** button, ⊕, to add classes.

5. By default, all the classes are assigned the **Value** of 0, the same **Color**, and **Custom color map entry** instead of **Label**. You can change them by double-clicking on a correspondent item. When you're done, your **Style** section will look similar to what is shown in the following screenshot:

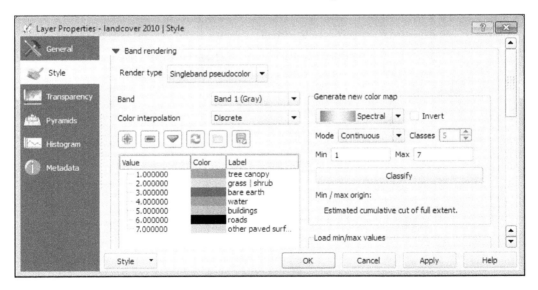

6. For the `height a.s.l., ft` layer, in the **Style** section, select **Render type** as **Singleband pseudocolor**.

7. The band will be set automatically, and in the case of smoothly changing values as in DEM, leave the default **Linear** option (under **Color interpolation**) unchangeable.

8. Now, we need to select an appropriate color ramp for the thematic content of the layer. A predefined ramps drop-down list doesn't show you all of the available diversity. To get this, select **New color ramp**. From the **Color ramp type** window drop-down list, select cpt-city. You will be shown all the available predefined color ramps, sorted into several thematic groups.

9. Click on the **Topography** group, select the **elevation** ramp (or any other ramp you like), and click on the **OK** button.

10. We will be asked to enter a name for new color ramp. Click on **OK** as we're satisfied with the existing one. Then, we'll be back to the main **Style** window.

11. Navigate to **Mode | Continuous**, and QGIS will create classes after you click on the **Classify** button. The **min / max** values range within which these classes will be created is defined in the **Load min/max values** section with the following options, as shown in the following screenshot:

    ° **Cumulative count cut**: By default, this is set to 2-98 percent of the data range and helps cut very low or very high data outliers. Choosing this type of data range setting, the initial image gains more contrast and better reflects differences in the data values.

    ° **Min / max**: The entire data range is taken into account, but the resulting map may look dull.

    ° **Mean +/- standard deviation ×**: Values within the given standard deviations (or deviations) define the data range.

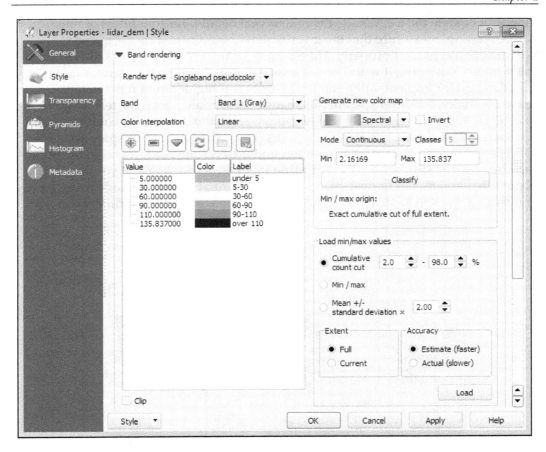

To load values according to the specified data range mode, choose **Actual (slower)** in the **Accuracy** section option and click on the **Load** button. Depending on the dataset, it may take some time, and when you see that the **Min** and **Max** values below the color ramp and **Mode** are updated, click on the **Classify** button. In the window to the left, you will see classes and their ranges set up automatically within the defined data range. This is as per the color ramp chosen. It is important to understand that every value represents the maximum limit of the class, and the highest value is not the real maximum value of the dataset, but the cumulative maximum count. You can adjust values by double-clicking on them and typing integer values, and clarify the legend by entering ranges instead of single values. Look at the preceding screenshot to understand these explanations.

Finally, let's style the `hillshade` raster to reveal details and roughness of relief. Before starting, make sure that the layer is above `height a.s.l., ft`, and if not, drag it and place it properly. Hill-shading simulates how sunrays illuminate the terrain. Combining it with DEM is a very popular approach in cartographic visualization for implementing semi-3D effects and highlighting relief details. Conventionally, semi-transparent hill-shading is overlaid above the terrain layer to achieve this. As a result, the final terrain visualization loses color contrasts and becomes dull. To overcome this, we will first move the **Global transparency** slider to 50 percent in the **Transparency** section under **Layer properties**, and in the **Style** section, select **Multiply** from the **Blending mode** drop-down list. If you're not sure about what exactly has happened, try to apply **Normal blending mode** using **Apply**, and then go back to **Multiply**. The following screenshot shows the difference:

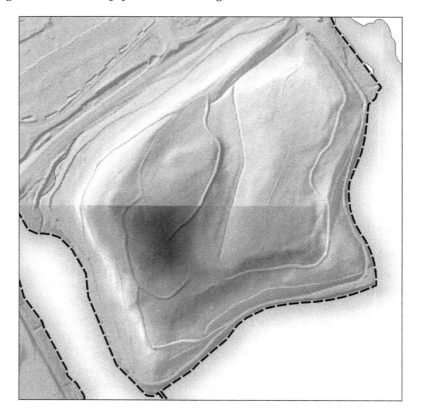

The preceding screenshot shows combined hillshade and DEM layers using simple 50 percent transparency (above) and the Multiply layer blending mode with 50 percent transparency (below).

Adding labels

Labeling is an important part of cartographic visualization. It significantly enhances data readability and understanding. Note that labeling is applicable only for vector layers because they contain landmarks and attributes (usually names) to be shown on the map. You can reach the **Labels** section from the **Layer Properties** dialog or by going to **Layer | Labeling** menu. Also, there is a **Label** toolbar that you can turn on or off by right-clicking on the toolbars panel contextual menu and use for fast access to the labeling options.

If you want to add labels to your layer quickly, just turn on the **Label this layer with** option, select the attribute field to be used for labeling from the drop-down list beside it, and click on **OK**. Labels will be added immediately to the layer. While this works well for personal and temporary use, there are many more labeling options available for the following label properties:

- **Text**: This defines the main properties of the text style, such as **Font**, **Size**, **Color**, **Type case**, and so on
- **Formatting**: This is used to organize and format labels as multiple lines
- **Buffer**: Text buffering defines properties of the buffer halo like size, color, transparency, and so on
- **Background**: They contain background options for the labels like shape, size, color of the background, and many more
- **Shadow**: These are shadowing options for labels and backgrounds
- **Placement**: These are advanced placement options for arranging labels and avoiding overlapping
- **Rendering**: These are options for labels and features for fast and clear label rendering

Now, we will go through some examples of layer labeling using advanced labeling options.

Labeling a point layer

In this example, we will label the `subway stations` point layer with station names and lines IDs. This means that in a label, we need to combine information from different fields, and to do this, we will use the `"name" || '\n' || "line"` expression.

Expressions give you advanced control over labeling options because with them, you can combine multiple fields, text, and functions to achieve the best result. The rules for constituting expressions are very simple: `field_name` is written in double quotes, and `text strings` in single. To merge them into a single expression, we use the `||` concatenation sign, and `\n` is used to start a new line.

1. From the **Font** drop-down list, choose for example **OpenSans** (or another desired font) and maximize its **Size** value to 9 points.

2. In the **Formatting** section, we only set **Alignment** to **Center**.

3. In the **Buffer** section, activate **Draw text buffer**. Set **Buffer size** to 1 mm and **Pen join style** to **Round**.

4. Don't forget to choose the color for buffering. Also, if you want to achieve more stylish effects, play with the **Transparency** slider and the **Blend** modes. We will not use any background options, so in this example, this section is omitted and we'll go right to the **Shadow**.

5. Activate the **Draw drop shadow** toggle. From the **Draw under** drop-down list, select **Lowest label component**.

If you want to see the result without leaving the dialog window, click on the **Apply** button from time to time.

6. In the **Offset** section, you can define the drop shadow angle. The value is entered manually or adjusted with the mouse arrow in a rotary switch beside.

7. Note that the shadow's outline depends on the label rotation angle. If you want to ignore it, the **Use global shadow** toggle should be activated.

8. Maximizing the **Blur radius** value softens the shadow, and you can achieve more complex effects using the **Transparency**, **Color**, and **Blend** modes.

9. By minimizing or maximizing the **Scale** value, the shadow can be made more subtle or obvious.

10. In the **Placement** section, activate the **Offset from point** mode and you will see the placement quadrants. Click on the lowest central quadrant button to place the label centered under the symbol marker. If the label is complex, as in our case, there is a chance that it will partially overlap the symbol. To avoid overlapping, use **Offset X,Y**, which are the values for the label's horizontal and vertical displacement. Enter a small positive **Y** value (for example, 3.0) to move the label slightly lower.

11. Finally, in the **Rendering** section, activate **Scale-based visibility** and enter `10000` in the denominator for the maximum possible scale. This means that our labels will be visible only within the 1:1 to 1:10000 scale range. If you want to prevent QGIS from hiding overlapped labels, activate **Show all labels for this layer (including colliding labels)**. Some labels may be rotated for better placement, or even the map canvas can be rotated. Then, you can decide whether or not to allow upside-down label rotation using **Show upside down labels**.

If the layer contains a lot of features, it is possible to use a number for **Limit number of features to be labeled to**. You can also use **Discourage labels from covering features**. As a result, your map will look neater. Depending on the labeling options you have chosen, your labeled layer may look similar to the following screenshot:

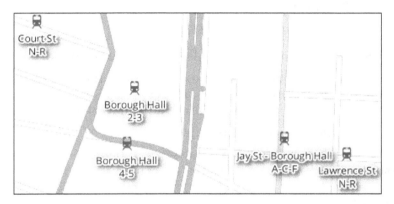

Labeling a line layer

The process of labeling a line layer is similar to the one described previously, but in the **Placement** section, you can find some line-specific options. The most important options are as follows:

- **Parallel**: The label is adjusted along a direction parallel to the major direction of the labeled line. This type of placement is good for conveying line curvature, but can miss small details.

- **Curved**: The label will be curved so as to reflect an original line's curvature. This is the best choice for labeling objects with complex geometry, such as rivers and pathways.

- **Horizontal**: Regardless of the line orientation, the label is always placed horizontally.

Labels can be placed above, on, or below line respectively. In the case of on-line placement, the line will be partially overlapped by the label. If you select several options at once, QGIS will define the best position and even take into account the line direction with **Line orientation dependent position** activated. The **Distance** option defines how far from the line the label will be placed (it is active only for **Above/ On line** positions). The maximum angle between curved characters defines how far you can bend a label. In the **Rendering** sections, the main line-specific options are located within **Feature options**. For example, here you can use **Merge connected lines to avoid duplicate labels**, which is very useful for working with a road network. With **Suppress labeling feature smaller than**, you can set up the values for small features to be ignored during labeling.

Labeling a polygon layer

In this section, we will add labels to the `zipocodes` layer:

1. Activate labeling and select the `zipcode` field to add labels.

2. In the text section, set **Font** to **Harlow Solid Italic** (or any other font you like) and enter `10` under **Size**. The labels are very simple, so we won't use any formatting, and instead of using buffer, we will work with **Background**.

3. Activate the **Draw background** toggle.

4. In the **Shape** drop-down list, there are several options available for defining the shape of the background. You can select from simple geometries (rectangle, square, ellipse, and so on) or more sophisticated shapes with SVG. Select **SVG**, navigate to the training `dataset/svg` folder, and select `plate.svg`. Set its type to **Buffer** and adjust **Size** to `1` mm. Also, if the SVG parameters are modifiable, you can select **Fill**, **Border color**, and **Border width**.

In the **Placement** section, you will see some polygon-specific options, as follows:

- **Offset from centroid**: Labels will be placed in the center of polygon within a selected quadrant and with a specified offset, if any. You can also specify exactly which centroid to use, visible or whole. If you select the visible polygon centroid, then the label visibility and placement will change dynamically while zooming and panning the map. If you select the whole polygon centroid, the labels will be static. Additionally, if you activate **Force point inside polygon**, the labels will tend to stay inside the polygon only.

- **Around centroid**: Labels will be placed in the center of the polygon within a specified distance.

- **Using Perimeter**: Labels will be placed along the border line (the **Above/ On/ Below** placement options are available, and it is possible to combine them to allow QGIS to define the best position). Activate **Line orientation dependent position** if you want take the line direction into account. The **Distance** value defines how far from the line the label will be placed, and the **Repeat** values vary the repetition frequency. This type of labeling is very useful for boundaries.

- **Horizontal (slow)**: All labels will be placed horizontally and will change their position dynamically with zooming and panning to stay inside the polygon. As QGIS constantly defines the optimal position, this type of labeling can work slower for large datasets.

- **Free (slow)**: QGIS will define the optimal position (including rotation) for the label within the current map view.

Try the different options yourself for a better understanding of label placement. The **Rendering** section is the same as for line layers, so you can avoid labeling some small features or limit their number.

Advanced labeling

Beside any labeling option, you can see the **Data defined override** button, ◀▤. It provides advanced control over labeling parameters. You can do this with an expression or by using a specially created attribute field that overrides the defined settings with custom parameters. Data-defined properties are very useful when you want to distinguish between labels for objects with different properties (for example, cities and towns with different populations) or apply your own placement positions.

Let's create three types of zipcode labels depending on population. The smaller label will be for areas where the population is less than 40000, the medium-sized label will be for a population of 40000-80000, and the biggest will be for a population of over 80000.

This means that we need to adjust the **Size** value of the font according to the `population` field values. From the **Data defined override** contextual menu, select **Edit...**, like this:

In **Expression string builder**, enter the expression shown in the following screenshot. In this expression, we use the population as the condition for font size value definition:

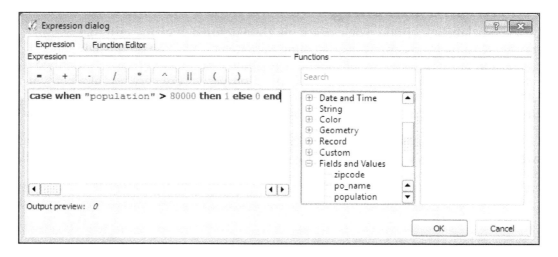

But what if we not only want to vary the size, but also want to underline the labels with the largest values? Then, we select **Edit** from the **Underline** button from the data override contextual menu and type the following expression: `case when "population" > 80000 then 1 else 0 end`.

If you are confused about exactly which values to enter for the parameter definition and how to combine them to get a meaningful result, select the **Description** item from **Data defined override**. A window, with a short explanation of the expected input, will be displayed. For example, from the following screenshot, which shows an explanation of the color input, we can conclude that if we want to change the label color according to the population, we can use `case when "population" > 80000 then '255,0,0,0' else '25,81,119,0' end`:

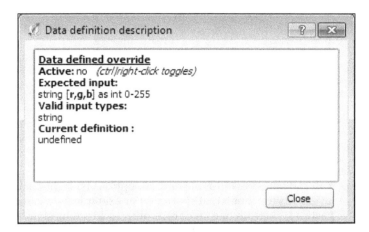

Managing styles

As you can see from the previous section, developing styles is a time-consuming task. But the good thing is that once developed, styles are not lost. They can be saved, applied to other layers, and imported from and exported to external sources.

The main style management options are available from the **Style** button at the bottom row of the **Style** section under **Properties** as shown. The menu is divided into sections separated by horizontal lines. The first section is responsible for loading and saving styles. The second and third are for managing multiple styles for the layer, and at the bottom is the toggle for activating different styles (inactive and grayed out by default, when there's only one style available).

 You can get access to some of these options by right-clicking on the layer's **Styles** contextual shortcut. For example, you can copy and paste a style from one layer to another, add a new style, or rename an existing style. Also, this is very convenient for rapid switching between multiple layer styles.

When you have finished polishing your style, it is wise to save it. There are three main options for doing this from the **Style** button of the **Save Style** menu:

- **QGIS Layer Style File**: The style is saved as a .qml file, which is a native QGIS format for storing styles.

- **SLD File**: The style is exported to a style layer descriptor (.sld) file. This file type converts the original symbology into a single-symbol or rule-based type. This means categorized, graduated, heatmap, and other types of symbology may not be supported properly. Renderer-based symbology may not be supported properly. It may be convenient to save the symbology in an .sld file if you plan to work on it within an external application, such as GeoServer.

- **Save in database**: We use this option to store and distribute all our data and styles within a single SpatiaLite database. When loading a style, it is important to provide a meaningful name and an exhaustive description. This is very handy because if you want somebody else to work with your data and style it properly, they just need to connect to the database, load the spatial layers and styles, and apply them.

Once the style is saved, you can use **Load from file (Style | Load Style | Load from file)** or **Load from database** to select and apply the style.

The QGIS community is very active in developing resources and eager to share them, so instead of spending a long time developing your own styles, you can apply ready styles provided by various users. We advise you to take a look at the following:

- Charley Glynn's OSM shapefile QGIS style sheets available at `https://github.com/charleyglynn/OSM-Shapefile-QGIS-stylesheets`
- 3liz styles for OpenStreetMap data in QGIS at `https://github.com/3liz/osm-in-qgis`
- Anita Graser's styles for SpatiaLite databases at `https://github.com/anitagraser/QGIS-resources/tree/master/qgis2/osm_spatialite`
- Ross McDonald's grayscale styles for OSM shapefiles in QGIS at `https://github.com/mixedbredie/OSM-Shapefile-QGIS-stylesheets/tree/master/QML%20files/greyscale`

You can download ready-to-use `.qml` files from there and apply them to your layers, but it's important that attributes of your layers be the same as those used in the ready styles. Otherwise, you can use these styles as basic templates and adjust the field names and values manually.

Using several styles for the same layer

You can also get quick access to some styling properties from the layer's contextual **Styles** right-click shortcut, as shown here:

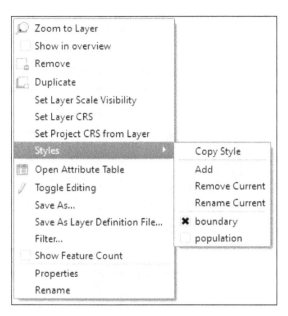

For example, you can easily copy a style from one layer and paste it to another. Moreover, you can apply several styles for one layer and switch between them when necessary. To add a one more style to the layer, follow these steps:

1. Click on the right-click contextual shortcut of **Add layer**, which is under **Styles**.

2. In the **New style** window, type the name for the new style, and click on **OK**, as shown in the following screenshot

3. The layer's appearance will not be changed, because it relies on the last applied style for now. Open the layer's **Properties** dialog and adjust the new style like this:

    ○   You can develop a new style in the **Style** section in a common way, as described earlier in the *Developing your own styles* section.

    ○   Also, it's possible to upload a ready style by going to **Style | Load Style**.

4. After the styling is done, click on the **OK** button. You will see that both the styles are available, either from the **Style** button menu or the **Styles** right-click contextual shortcut.

5. You can develop as many styles as you wish, and use the toggle beside their names to switch between styles. Also, you can use the **Add**, **Remove Current**, and **Rename Current** shortcuts to manage multiple styles.

Adding base maps

Base maps are ready-to-use background maps that provide contextual and spatial information additional to your data. They can be satellite imagery, general maps from various sources, or even self-prepared custom maps. In this section, we will take a look at the most popular types of background maps and how to load them into QGIS to combine them with your data.

The OpenLayers plugin

This is one of the most popular QGIS plugins, as it allows simple addition of base maps from numerous popular map providers (OpenStreetMap, Google Maps, Bing Maps, and so on). Install the plugin as described in *Chapter 1, Handling Your Data*, and make sure that it's active after installation.

Loading base maps is simple; go to **Web | OpenLayers**, select the provider, and click on the map you want to add. The map will be loaded into the map canvas and appear in the **Layers** panel. By default, layers are added to the first layer group, but you can drag and drop them wherever you want. The layer can be shown or hidden with the toggle beside its name, and deleted from the project with the **Remove** right-click contextual shortcut.

You can gain extended control over layers and navigation by activating the **OpenLayers Overview** panel from the menu. The panel will appear in the bottom-left corner under the **Layers** panel.

Activate the **Enable map** toggle and select a map for overview from the drop-down list. You can use two different maps in the overview window and the map canvas window for comparison. If you want to load the map into the canvas, click on the **Add map** button beside the drop-down list. For simpler navigation, there is a red cross in the main map window. It marks the center of the overview extent. You can hide it by clicking on the corresponding toggle. Also, the overview map can be saved as a .jpeg image, or a rectangular extent can be copied to the clipboard as KML.

The **OpenLayers** plugin is very useful because of its numerous maps and simplicity, but there are a number of constraints you should pay attention to while working with it. First of all, this plugin is aimed at providing base maps only, and before using it to perform other tasks, it is highly recommended to study the provider's licensing terms. Secondly, note that adding any layer from the plugin's list automatically changes the original map projection to **EPSG: 3857 WGS 84 / Pseudo Mercator**.

This is because the plugin fetches data originally provided in EPSG: 3857, and instead of reprojecting, it suppresses map projection and automatically reprojects the user's data. Last but not least is the fact that with the OpenLayers plugin, you cannot rely on scale and map measurements. This is because the EPSG: 3857 WGS 84 / Pseudo Mercator projection it uses was designed not to minimize object distortions (shape, area, distance, and so on) but to fit the entire globe in such a way that it could be shown on the web map. All measurements in this projection are performed on a sphere, and will most likely be much larger than expected. In a few words, this projection is good for visual exploration but not for measurements. To overcome these limitations, you can use other approaches.

Adding WMS/WMTS layers

The **Web Map Service / Web Map Tile Service (WMS/WMTS)** is a popular web-protocol for spatial information transferring To add WMS/WMTS base layer to your map perform the following steps:

1. To load data from web services, go to **Layer | Add Layer | Add WMS/WMTS**, use the corresponding button in the **Mange layers** toolbar, or use the *Ctrl + Shift + W* keyboard shortcut.

2. In the **Add layer(s) from a WM(T)S Server** window, click on the **New** button to configure new connection parameters.

3. In **Create a new WMS Connection**, enter connection details. Type **Name** and **URL** for the connection and authentication parameters, if any. After filling in these details, click on **OK**. A newly created connection will show up in the drop-down list under the **Layers** tab.

4. Click on the **Connect** button to get information about the available layers, select the layer you want to use, click on **Add**, and then click on **Close** to leave the window.

In the case of using WMS-layers, the data will automatically be reprojected. Also, you will have access to the layer's **Properties** dialog and styling parameters, such as transparency, blending, and color mode available for modification. Of course, these layers are available for creating high-resolution printing maps.

Adding TMS layers

The **Tile Map Service** (**TMS**) is another way of providing spatial data through the Internet in the form of georeferenced images (tiles). To load TMS data into QGIS, install and activate the **TileMapScale** plugin, as described in *Chapter 1, Handling Your Data*. After installation, the plugin is available from the **TileMapScale** menu under **Plugins**. The plugin's panel consists of two tabs. On the first tab, **Tool**, you can see a drop-down list with the available TMS datasets. With the **activate zoomlevels** toggle, scales will be set automatically to fit the tile's zoom level. Under the **Options** tab, you can activate **Use 'On-The-Fly' Transformation** if you want the tile layers to be adjusted to map projection. Set **min/max Zoomlevels** and click on the **DPI** button to set the resolution directly.

> The plugin stores dataset descriptions in the UserName\ .qgis2\ python\ plugins\ TileMapScaleLevels\ datasets folder as .xml files. These files use the GDAL TMS minidriver format to describe the data source parameters. You can read more about this at http:// www.gdal.org/frmt_wms.html. Using the examples provided with the original plugin installation and by studying the documentation, you can create .xml files for your own data providers.

Other than this plugin, you can also use the **TileLayer** and **QuickMapServices** plugins to work with TMS.

Summary

Now you know everything necessary for organizing and styling your data. You are also able to add informative labels, save your styles for future use and sharing, and provide some spatial background context with base maps from various sources.

The next step is to prepare your maps for printing.

3
Presenting Data on a Print Map

Print maps are a great tool for sharing your spatial information because they can be included in presentations, publications, and reports as high-quality graphics or printed in a conventional way. QGIS has powerful capabilities of creating single-page or multipage (so-called atlases) maps that you can export in multiple, easy-to-distribute, and print graphical formats (for example, .pdf, .png, .svg, and so on).

In this chapter, you will be exposed to QGIS's main tool for creating print maps, called **print composer**, and will learn how to use it to get outstanding results.

Print composer

With print composer, map designing becomes easy and intuitive, as it allows you to set up a layout and add a map and its necessary elements, such as text labels, the legend, the scale bar, and the north arrow. Moreover, you can significantly enliven your map by combining multiple overviews, images, drawings, and HTML labels. To simplify map generation and save your time, you can use map templates and the atlas generator functionality.

To enter print composer mode, go to **Project** | **New Print Composer** from the menu, click on the ▢ button on the **File** toolbar, or just use the *Ctrl + P* keyboard shortcut. In the **Composer title** dialog window, you will be asked to give a title to your map. Click on the **OK** button for the title to be generated automatically (Composer 1, Composer 2, Composer 3, and so on; you will be able to rename them later if you want).

In the upper row of the print composer window, there are drop-down menus that provide access to its options and functionality. The upper row of toolbars provides quick-access buttons for the same functionality. Note that the **Composer Items** toolbar is, by default, placed separately on the left side of the window because it provides quick access to all elements that are usually necessary to constitute a map, for example, a map canvas, a legend, a scale bar, labels, annotations, and so on. In the central part of the window, you can see an empty page that represents the composer canvas available for item arrangement, as shown here:

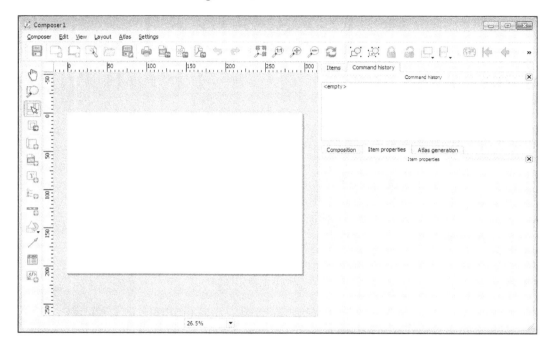

In the right part of the print composer window, there are two panels with tabs. The upper panel holds the following tabs:

- **Items**: This displays the list of map items added to the map. It allows you to define their visibility, rename, reorder, and lock them for changes.

- **Command history**: This displays the list of commands used. With the history, you can easily roll back changes and have better control over the process of map generation.

The lower panel holds three tabs, as follows:

- **Composition**: This is responsible for **Paper and quality** (page size, number, and resolution) and **Guides and Grid** (spacing and snapping tolerance) so as to simplify map item placement.

- **Item properties**: This is a dynamic tab whose contents changes automatically, depending on the map item currently selected.

- **Atlas generation**: Once the **Generate an atlas** toggle is active, you can set and modify its parameters.

> Note that it is only the default panels' placement that can be rearranged. In order to change the panel/tab position, hover your mouse arrow over its perforated area, which looks like this:
>
> | Items | ✖ |
> |---|---|
>
> Now, drag and drop it at another place. You can also close any unnecessary panel by clicking on the button in its top-right corner. Also, panels can be turned on or off with the toggles that become available after right-clicking on the menu or the toolbar's panel, that is, when you see something like this:
>
> ✖ Command history
> ✖ Atlas generation
> ✖ Item properties
> ✖ Composition
> ✖ Items
> ✖ Composer
> ✖ Paper Navigation
> ✖ Composer Item Actions
> ✖ Composer Items
> ✖ Atlas

You can get additional control over composer settings from its **Composer Options** menu, which can be found under **Settings**. For example, you might want to change the **Default font** that will be used for the legend and text labels. For further work in the **Grid appearance** section, we set **Grid style** to **Solid** and leave **Grid and guide defaults** unchanged, as we will configure them later.

The initial setup – page format and other essentials

Before creating the map, we should set up our working space, which is mainly defined by page settings. All the necessary options are present under the **Composition** tab. Let's take a closer look at the **Paper and quality** section. The default page size in the **Presets** drop-down list is **A4**, but you can select from among 23 other sizes. We are going to create our own page by performing the following steps:

1. Select **Custom** from the list.

2. The **Width** and **Height** fields will be activated. Enter a value of 200 for both of them (this means that we are going to create a square page for a map composition). The value depends on the units chosen in **Units**, and by default, it is set to **mm** (millimeters).

As we are going to create only one map, we leave **Number of pages** as 1, but you can always enter more if you're going to combine several pages into one document. This is because you could have several maps in a single-page composition. As shown in the example, we're using a square page. So, it doesn't matter whether it has the **Portrait** or **Landscape** page orientation. **Page background** is a very important item as it allows you to achieve cool cartographic visualization effects. By clicking on the **Change** button, you can see that, by default, it is set to **Simple fill** with white color (which we are going to use). However, you can change it and select from among multiple filling options to create an original filling pattern.

> For example, you can find nice raster or SVG textures and use them as background patterns by selecting **Raster image fill** or **SVG fill** respectively. Otherwise, you can select **Gradient** or **Shapeburst fill** to play around with some color grading effects.

Export resolution defines the quality of the exported map in dpi units. The higher this value, the better the map. However, the resulting file will also be bigger. For common purposes, such as publications, reports, and presentation graphics, 300 dpi is enough. When the **Print as raster** toggle is active, the entire composition is converted into raster before exporting to .pdf. The **World file on** toggle allows you to attach a .wld file that contains information about a geographic reference to the image export output. This means that your map can not only be viewed in graphical or preview software, but can also be opened and aligned properly directly in GIS software.

Now, we will adjust the grid that will be used to align map items within a map layout:

1. In the **Guides and Grid** section, enter 5 mm as the **Grid spacing** value to make it denser, and set **Snap tolerance** to 10 pixels (the higher the value, the stronger the snapping).

2. To show the grid, activate the **Show grid** button, which is under **View**, or use the *Ctrl + '* keyboard shortcut.

3. Toggle **Snap to grid**, which is under **View**, or pressing *Ctrl + Shift + '* activates snapping.

Instead of using a predefined grid, you can create your own guidelines by placing the mouse arrow over horizontal (for vertical guidelines) or vertical (for horizontal guidelines) ruler bars and clicking and dragging them along the page canvas. Deactivate **Snap to Grid** and use various options from the **View** menu such as **Show/Clear** or **Snap to Guides**. The **Smart Guides** option helps you detect the best placement position automatically based on the other items in your composition.

Adding and customizing a map

In this section, we are going to create a population map of the Brooklyn borough. Make sure that in the main window, the following layers are active: `Brooklyn borough boundaries`, `zipcode` (with the population style), `NY borough boundaries`, and `water area`.

To add a map to your print layout, click on the **Add new map** button in the **Composer Items** toolbar, and drag the mouse arrow diagonally over the page while holding down the left mouse button. You will see a rectangle drawn on the page and a map inside it.

Note that the content of the map changes dynamically to reflect the content of the map canvas in the main window. This means that if you turn on/off some layers in the main window, or change their styling, the map in the print composer window will be changed too. If you don't see the changes, click on the **Refresh view** button to update the map. We will discuss how to suppress this behavior soon.

The map frame can be moved or resized by dragging the special squared markers on the borders' centers and corners. If you don't see the frame and markers, it means that the current item is inactive. To activate it, click on the **Select/Move item** button in the **Composer Items** toolbar. If you want to move a map inside the frame, use the **Move item content** button, in the **Composer Items** toolbar. Also, as long as this button is active, you can zoom into the map with the mouse wheel (or with *Ctrl* and the mouse wheel). Using markers, extend the map in such a way that it covers the whole page.

After placing the map properly, we can proceed to the advanced options on the **Item properties** tab. The tab brings together several sections that can be expanded or collapsed by clicking on the ▶ symbol beside them.

The **Main properties** section allows you to select from the following map preview modes:

- **Cache**: This renders the map in the current screen resolution, which will not change while zooming in/out, but the image itself will be scaled properly.

- **Render**: The map resolution will be adjusted to the maximum value while zooming. This mode gives better a preview picture, but is slower comparing to **Cache**.

- **Rectangle**: Instead of a map, a `Map will be printed here` message will be shown in an empty box.

When you enter some values in the **Scale** field, the map extent changes dynamically. Set its value to `100000` and use the **Move item content** button to locate map within frame properly. The following screenshot shows you how the tab **Item properties** looks like at this stage:

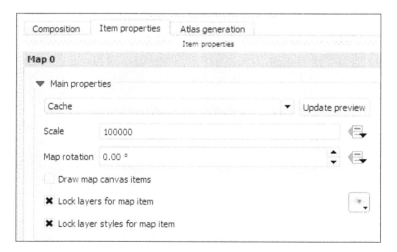

With the **Map rotation** field, you can define an angle in degrees to rotate the map clockwise within its frame. Note that the angle of rotation will be applied automatically to the grid lines to adjust them properly with respect to the map's content. There are three toggles and they provide the following options:

- When the **Draw map canvas items** toggle is activated, existing annotations from the main map canvas window will be added to the map.

- The **Lock layers for map item** option is very useful if you don't want your print map to reflect changes made in the main map canvas window. So, the layers will remain the same, but their styling and labeling will change according to the main window. We activate this option because we don't want the composition of layers to be changed.

- **Lock layer styles for map item** blocks the map item from style changes made in the main map canvas window. Activate this option too. Jointly applied with the previous option, it allows us to block the composer map from changes made in the main map canvas window (that is, the layers' style and composition).

On the right-hand side of the toggles, there is a **Set layer list from a visibility preset** button, . This button allows us to quickly define map item content from the visibility presets on the toolbar of the **Layers** panel. To load a preset, click on the small triangular arrow in the bottom-right corner of the button, and select the preset you want to show on the map. The map's content will be changed according to the preset and the layers will be locked automatically.

The **Extents** section provides various options to define and adjust the map item's extent:

- Manual definition, by entering **X** and **Y** values in the correspondent fields.

- **Set to map canvas extent**: The current map extent will become the same as the extent of the main map canvas window.

- **View extent in map canvas**: This does exactly the opposite of the previous button. It changes the map canvas extent in the main window according to the current map item extent.

The **Controlled by atlas** section will be covered later, when we talk about atlas generation in the *Creating atlases* section.

The **Grids** section allows you to add coordinate grids and frames to the map frame. To add a new grid, click on the button and adjust its main properties, such as **Grid type**, **Interval**, **Grid frame**, **Coordinates**, and so on.

You can combine multiple grids that use different coordinate reference systems within a single map item. Just add one more grid. Then, in the **Draw "Grid 2" grid** section, click on the **Change** button beside **CRS** and select the CRS you want to use. A common use case of this option is to combine a projected CRS, which uses linear measurement units (feet and meter), with a geodetic CRS, which provides latitude and longitude coordinates in degrees.

The **Overviews** section allows you to combine multiple overviews in various scales into a single map composition. This will be covered later, in the *Working with map overviews* section.

The following sections are common for various map items:

- The **Position and size** section allows you to define the size of the item frame in page units. **Reference point** decides which corner of the item is specified by the X and Y positions.
- The **Rotation field** values set the item's rotation angle in degrees.
- The **Frame** toggle enables the frame around the item, and provides access to its options, such as **Frame color**, **Thickness**, and **Join style**.
- **Background** is responsible for the frame's fill color.
- **Item ID** can be used with web clients to create links between map items, and also for easier identification of specific items within the **Items** panel.
- **Rendering** provides access to the **Blending mode** and **Transparency** options that allow us to apply professional graphics affects.

Adding and customizing a legend

A legend is a necessary element in any map. It provides explanations of the layers' symbology and helps us read and understand the map. To add a legend, click on the **Add a new legend** button in the **Composer Items** toolbar, and drag the mouse arrow diagonally over the page while holding down the left mouse button. You will see a rectangle drawn on the page and a legend inside it. By default, all layers in the project will be included in the legend, so it may look too big to fit the layout. With the legend item selected, you can use the **Item properties** tab to customize its content and appearance. There are several legend-specific sections that should be discussed.

The **Main properties** section allows you to customize the following options:

- **Title**: By default, the `Legend` title is used, but you can type any other text here, or leave the field empty to remove the title.

- **Title alignment**: This can align the title to **Left**, **Center**, or **Right**.

- **Map**: By default, there is only **Map 0** available, but if you combine several maps within a single page layout, you can decide which map the legend will refer to.

- **Wrap text on**: Type here a symbol that will be used to separate long lines into shorter chunks of text. Later, you can edit text in the legend item and insert this symbol wherever necessary to forcibly start a new line.

The **Legend items** section is responsible for the content of the legend. By default, the **Auto update** mode is active, and all available layers are loaded into the legend. Turn the **Auto update** mode off and the buttons below the legend contents tree will be activated, providing the following options (from left to right):

- The **down** ▽ and **up** △ buttons are used to move items and change their order. Otherwise, you can simply drag and drop an item at a new position. After any changes the legend's item content is immediately updated.

- The **add group** button allows you to add new groups and subgroups to develop a legend hierarchy.

- The **plus** and **minus** buttons are responsible for adding and removing any item, whether it is a group or a layer. Note that only layers active in the main map canvas window are available for adding.

- The **edit** button allows us to change the text for any selected item (group, subgroup, layer, or symbology label). For example, if you want a long line to be wrapped, insert in the text symbol defined in the **Main properties** section. These changes are applied only within a **print composer legend** item and don't affect map items in the main window.

- The **sum** $\sum$ button shows the feature count for each class of the vector layer.

- The **filter** button filters the legend by map content; that is, only the legend items (layers and classes) that are shown in the map item will be included in the legend.

In the **Fonts** section, you can define a common color and different fonts for the title, subgroups, groups, and items to enhance the legend's readability. If the legend contains many items and is too long to be shown in a single column, it can be split in the **Columns** section. Here, you can define the number of columns in the **Count** field. By default, the column's width is adjusted according to the content. If you want it to be set equally for all columns, activate the **Equal column widths** toggle. If the **Split layers** mode is on, categorized layer symbology items will be allocated equally among columns; otherwise, categories are not allowed to be moved into another column. In the following screenshot, you can see how this works. At the top, there is a legend in which layer splitting is prohibited, and at the bottom, the categories are equally split between three columns.

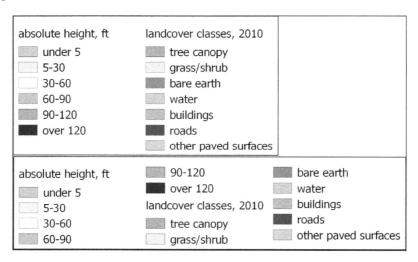

In the **Symbol** section, the default **Width** and **Height** values can be modified manually. **WMS Legend Graphic** is used to add a legend for a WMS layer, but this is if it is supported and provided by the WMS server. In the dialog window, you can resize the legend width and height that is provided as a raster image. The **Spacing** section allows you to refine the legend's layout and readability by adjusting the spaces between its various items: title, groups, subgroups, symbols, icon labels, bounding box, and columns.

Other sections, such as **Position and Size**, **Rotation**, **Frame**, **Background**, **Item ID**, and **Rendering**, provide the same options as those for the map item.

Other map items

For now, our map contains two necessary items: the map itself and the legend that explains its content. However, there are more items that can be used to improve its aesthetic view and geographic readability, such as a scale bar, a north arrow, text labels, and so on.

Scale bar

Before adding a scale bar item, make sure that your map canvas is using the projected coordinate reference system in linear measurement units (meters and feet). To add a scale bar, click on the **Add new scalebar** button in the **Composer Items** toolbar, and drag the mouse arrow over the page while holding down the left mouse button. You will see a rectangle drawn on the page and a scale bar inside it. With the scale bar item selected, use the **Item properties** tab to customize its appearance.

In the **Main properties** section, select a map to associate the scale bar with. As you will be working with only one map, **Map 0** will be set by default. The **Style** drop-down list defines the appearance of the scale bar:

- **Single** or **Double Box** is responsible for scalebar appearance as a singular or double zebra box respectively

- **Line Ticks Middle**, **Down**, or **Up** produce a simple horizontal line with tick marks on it for fixing scale units

- **Numeric** displays the scale as a numeric ratio

A graphical scale bar is the right approach to represent scales on maps that appear as images in presentations, reports, and publications. It is advisable to use it instead of a numeric ratio scale because the graphical bar will be proportionally scaled during image resizing. If you still need to use a ratio scale, make sure that the image will be distributed in its original size.

For the purpose of this tutorial, select **Double Box Style**. The **Units** section is responsible for the right representation of measurement units:

1. From the drop-down list, select the **Feet** projection measurement units. The scale bar will be converted into a bar that measures miles. Similarly, conversions are supported by **Meters** and **Nautical Miles**. **Map units** can use the original map units, but you will have to specify them.

2. In the **Label** field, type in mi to use the abbreviation of measurement units (miles, in our case).

3. **Map units per bar unit** represents the ratio of map units to bar units. Enter the value as 5280 here, as we know that 1 mile is equal to 5,280 feet.

Don't worry about strange, collapsed appearance of your scale bar; we will fix it in the **Segments** section as explained:

1. In the **Size** section, you should enter the value to define how long a segment will be. If you want it to be 1 mile long, enter 5280; to make it 2 miles long, enter 10560 (*5,280×2*); and so on. The principle is simple—just use the **Map units per bar unit** ratio that you entered in the **Main properties** section and multiply it by the number of scale bar units that should be in one segment. For the purpose of this tutorial, we will make one segment equal to 1 mile and enter a value of 5280.

2. In **Segments**, enter value 2 for **left** and 1 for **right**.

3. Increase the scale bar's **Height** value to 5 mm.

In the **Display** section, you can customize the following scale bar properties:

- **Box margin**: The bigger this value, the longer the distance between the scale bar item 's contents and the border of its bounding box.

- **Labels margin**: This defines the space between the scale bar and text labels.

- **Line width**: This defines the width of the scale bar's outline.

- **Join style**: This defines the appearance of the scale bar's corner. It is available only for box type scale bars.

- **Cap style**: This defines the appearance of line endings. It is available only for line type scale bars.

- **Alignment**: This aligns the scale bar and its labels within the bounding box. It is available only for the numeric scale type.

The **Fonts and colors** section provides access to the eponymous functionality. To change the font or color, click on the button and select the necessary item from the dialog window. Note that **Fill color** and **Secondary fill color** work only with box type scale bars. In the following screenshot, you can see the example of a simple **Double Box** scale bar with two segments on the left side and one on the right side:

Sections such as **Position and Size**, **Rotation**, **Frame**, **Background**, **Item ID**, and **Rendering** provide the same options as those that were covered for the map item.

North arrow

To add a north arrow, click on the **Add image** button in the **Composer Items** toolbar, and drag the mouse arrow over the page while holding down the left mouse button. You will see an empty rectangle drawn on the page. With **Image item** selected, use the **Item properties** tab to customize its appearance.

First of all, you should navigate to the **Image source** file you want to use:

1. Expand the **Search directories** section. In this section, you can set up **Image search paths**. By default, you will see the `QGIS installation path\qgis\svg\` path and its subfolders. You can always redefine paths with the **Remove** and **Add** buttons, and get access to your own images.

2. In the **Loading previews...** window, you will see thumbnails of the available images as shown in the following screenshot. Clicking on them automatically loads them into the image item frame. For the purpose of this tutorial, we select `NorthArrow_02.svg`.

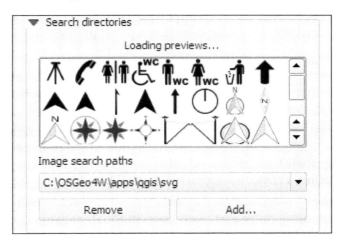

Now that you have selected the north arrow, you can adjust some of its **Main properties**. **Resize mode** defines the approach to image adjustment within its frame:

- **Zoom**: The image will be adjusted to the frame and its proportions will be saved.

- **Stretch**: The image will be stretched within the frame, ignoring its proportions.

- **Clip**: The image will be resized to its original size, and the frame will be used to clip the image, so only the part of it will be visible. This mode is appropriate for raster images only.

- **Zoom and resize frame**: Both the image and the frame will be resized to fit each other. First, the image will be proportionally adjusted to fit the frame, and then the frame will be adjusted to encompass the image.

- **Resize frame to image size**: The frame will be adjusted to the original image size.

For the purpose of this tutorial, we will use the **Zoom and resize frame** mode. In this mode, the **Placement** option is inactive, but you can use it when selecting the **Zoom** and **Clip** modes to position an image within its frame. Similarly, **Image rotation** is also available for the **Zoom** and **Zoom and resize frame** modes. Activate the **Sync with map** toggle if you want to rotate the north arrow synchronously with the map.

Other sections, such as **Position and Size**, **Rotation**, **Frame**, **Background**, **Item ID**, and **Rendering**, provide the same options as those covered for the map item.

Other items

So far, we have covered only the most important and the necessary map items. You have probably noticed that the **Composer Items** panel provides access to many more items, as follows:

- **Label item** T: This provides control over text (or HTML) labels and their various properties, such as **Font**, **Font color**, **Vertical** and **Horizontal** (alignment), **Position and Size**, **Rotation**, and so on.

- **Shape item** : This allows you to add simple geometric figures, such as rectangles, triangles, or ellipses, to the layout. In the **Item properties** tab, you can change their type, style, and other basic options, such as **Position and Size**, **Rotation**, **Frame**, **Background**, **Item ID**, and **Rendering**.

- **Arrow item** : This is a very useful element for highlighting relations between multiple maps and overviews. In the **Item properties** tab, you can select **Line style**, **Arrow outline** and **fill colors**, **Arrow outline** and **width**, and **Start** and **End markers**.

- **Attribute table item** : This item adds a vector layer attribute table to the map layout. If you want to do so, select **Layer** under **Item properties** tab. By clicking on the **Attributes** button, you will be taken to the **Select attributes** window, where you can modify the table's structure and appearance by adding or removing unnecessary columns; changing headings, column alignment, and width; and applying sorting options.

For this item, you can also apply **Feature filtering** to exclude unnecessary or excess elements. Within the **Appearance** and **Font and text styling** sections, you can find various options for enhancing table readability.

- **HTML frame item** : This is responsible for adding a web page or user-generated content using HTML markup. For example, you can type the following expression in the **Source** window: `Map data is provided by <a href='https://nycopendata.socrata.com/'>NYC Open Data Portal</a>`. This adds a link to the original website.

When everything is ready, you can export your map from the menu, by navigating to **Composer | Export as Image** or **Export as SVG** or **Export as PDF**, or by clicking on the relevant buttons in the **Composer** toolbar, where is responsible for image exporting, is for exporting as SVG, and for exporting as PDF.

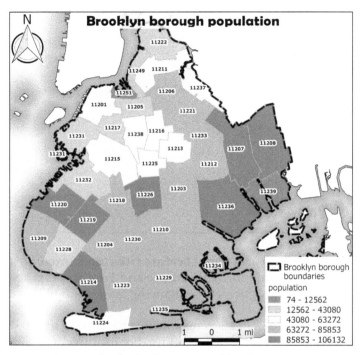

An example of a simple map created in print composer

Working with multiple maps and overviews

Sometimes, there is a need to combine multiple maps within a single layout. For example, suppose you need to combine a few different maps on one page, or show the location of an area of interest within a general region. While these tasks look similar, they require slightly different solutions, which we will explore in the following sections.

Combining several maps within a single print composer

For our example, we will combine two maps—relief and hurricane evacuation zones—and do this within a single print layout:

1. In the main window, activate the layers that will be used in the relief map— `Brooklyn borough boundaries`, `NY borough boundaries`, `water area` (to show the shoreline and water bodies properly), and finally, `hillshade` and `height a.s.l., ft`.

2. Then, open the **New Print Composer** window, either from the **Project** menu or using the *Ctrl + Shift + P* keyboard shortcut. Adjust **Page Size** for this example to **A3**, set **Grid Spacing** to 5 mm, and activate the **Show Grid** and **Snap to Grid** options from the **View** menu.

3. Add a new map item and adjust its size to 200×200 mm in the **Position and Size** section of the **Item properties** tab. Set the scale to 100000 and place the map content properly. After that, activate the following toggles to prevent changes to the map's content:

    ○ **Lock layers for map item**
    ○ **Lock layers styles for map item**

4. Add and adjust another necessary map items, such as the scale bar, legend, north arrow, and so on. Compose a legend, removing unnecessary layers. If you want to block some items from changes, lock them in the **Items** panel, as shown here:

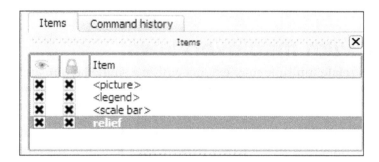

5. When you are done with the relief map and its elements, go to the main map canvas window and activate the layer you want to show on the map. For example, activate `hurricane evacuation zones` and `primary residential zoning`, and deactivate `hillshade` and `height a.s.l., ft`.

6. Add a new map item and repeat steps 3 and 4. After you are done, the result may look like what is shown in the following screenshot:

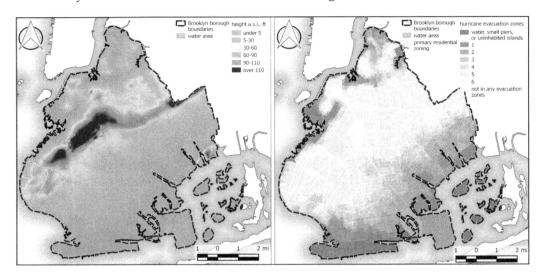

You can combine any number of maps within a single layout, and use different scales and individual legends for them. The only thing you should always remember is to use **Lock layers (styles) for map item** to prevent any changes within a map.

This approach is simple and effective when you need to present similar maps within a single page, but sometimes there is a need to show different areas of interest in different zoom levels, and also relate their extents with each other. In such cases, you should use so-called map overviews.

Working with map overviews

Map overviews relate multiple map extents within a single layout, and they are very useful for showing how an area of interest is related to a bigger spatial context, for example, a region within a country, a district within a city area, and so on.

In this example, we will combine three overviews into a single map layout to show how Brooklyn is related with the whole of New York and how the area of interest within Brooklyn is related to the borough's extent. To achieve this, you need to perform the following steps:

1. In the main map canvas window, activate the NY borough boundaries layer and deactivate all other layers. Create a new print composer, use the default **A4** page size, and adjust any other properties required. Add a new map item to the layout. This map will be used to show simple boundaries and provide spatial context for the main map, so it may be encompassed in a small 95 x 95 mm frame with a scale of 1:515,000. After you have adjusted the map view, don't forget to select **Lock layers for map item** and **Lock layers styles for map item**.

2. Now, you will be adding the main map, and it will be the core of the layout. In the main map canvas window, activate the layers you wish to show on the map (for example, hurricane evacuation zones and primary residential zoning). Again, add a new map item to the layout and adjust its size, scale, and position. For example, it might be of 200 x 200 mm with a scale of 1:97,000. Again, use **Lock layers for map item** and **Lock layers styles for map item**.

3. Finally, add the third map. Activate the layers you wish to show in the map in the main map canvas window (for example, hillshade, and height a.s.l., ft). Add a new map item. Adjust the map size to 95 x 95 mm and set the scale to 20000. Move the item content to show the area you are interested in and select **Lock layers for map item**.

4. Now, we need to relate the map extents with each other. We can do this with the help of overviews. Activate the very first map item you added and expand the **Overviews** section under the **Item properties** tab. To add a new overview, click on the 🔘 button, and to show it on the map, activate the **Draw "Name" overview** toggle. To relate an overview with a map item extent, select the map from the **Map frame** drop-down list. You can change the default **Frame style** using the **Symbol selector** dialog window.

Blending mode is useful for achieving more stylish effects when combining several overview frames. The **Invert overview** toggle will invert the fill of the overview outside of its area. **Center on overview** will move the map item content to place an overview in its center.

Add another overview that encompasses **Map frame** with a smaller extent using the button. You can combine as many overviews as you want. Use the ▬ button to remove unnecessary overviews and the ▲ and ▼ buttons to reorder them.

5. Go to another map and add an overview to it. Select the map item you are interested in and go to its **Properties** tab. Expand the **Overviews** section and click on the ⊕ button to add an overview. Select a necessary map from the **Map frame** drop-down list and adjust the overview's **Frame style**.

> QGIS makes it possible to combine multiple overviews and produce complex results. To highlight the differences and relations between different overviews, use individual map frames and scale bars.

In the following screenshot, you can see the resulting map. It combines three map items and relates their extents with overviews.

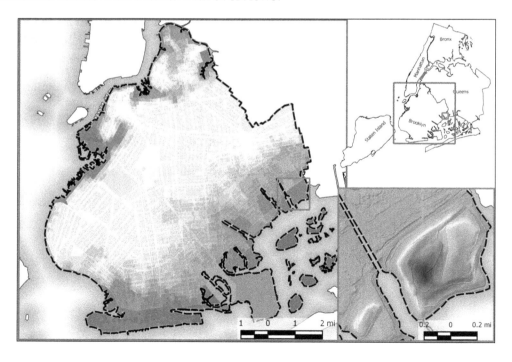

Creating atlases

In QGIS, an atlas is a multipage map created in an automated way with the atlas generation functionality. The idea underlying atlases is very simple—very often, the user needs to generate several small maps, instead of one big map. To arrange multiple maps properly within several pages, an atlas coverage layer is used. This coverage layer may contain various types of geometry (points, lines, and polygons) that are used to define the general extent of the map, separate sheet's extents, their quantity, and their sequence. The most common example of an atlas is a large area (for example, a country, borough, or watershed basin) separated into several parts based on its subunits (for example, regions, states, counties, or subbasins).

In this example tutorial, we will generate multiple street network maps using the `zipcode` layer elements as bounding extents for multiple maps. Moreover, we will use a specially designed plugin to mask all excluded areas and achieve stylish cartographic effects:

1. First of all, we enable the layers that we are going to create our map from: `Brooklyn borough boundaries`, `zipcode`, `water area`, `parks`, and `roads`.

2. For the purpose of this tutorial, we are going to use the external **Mask** plugin. This plugin allows the user to create a map-masking layer and also filter the labeling of other layers, removing labels that are outside the area of interest. It is also able to interact with the atlas generation process. Install and activate the **Mask** plugin as described in *Chapter 1, Handling Your Data*.

3. When the plugin is activated, we need to provide it with layer features on which masking will be based. In our case, this will be `zipcode`. Activate it. Then, using the **Select features by area or single click** mode, click on the map canvas to select any feature.

4. Once the selected polygon is highlighted, go to **Plugins | Mask | Create a mask** and open the **Create a mask** dialog window. In this window, you can define the following properties of the mask:

   ° **Style** of the masking polygon, which, by default, uses **Inverted polygons renderer** with the **Shapeburst fill** option, modified by **Transparency**.

   ° The **Buffer** toggle generates a buffer around the area of interest whose width and complexity are defined by the **Units** and **Segments** values.

   ° On-the-fly simplification is used for the mask geometry used in expressions. The higher the value, the coarser the geometry.

   ° **Function used for labeling filtering on polygons** defines how exactly labels outside the masking polygon will be eliminated.

- ◦ **Function used for labeling filtering on lines** defines the exclusion mode for line labels. Select **The mask geometry contains the line** from the drop-down list. In the available layers list, activate roads (note that this layer is available only when its labeling is turned on).

- ◦ Activate the **Save as** toggle and navigate to the catalog where you want to save the newly created mask layer. Otherwise, a so-called memory (temporary) layer will be created, and it will not be saved after closing the project.

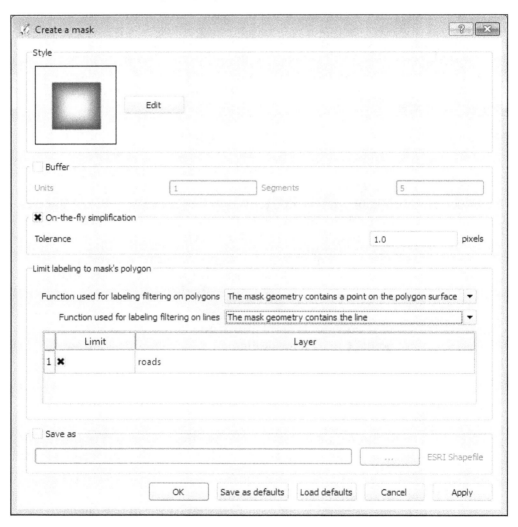

5. After you've clicked on **OK**, the mask layer will appear in the layers panel, and you will see its fancy styling effect applied to the map canvas. The selected polygon feature looks highlighted, while the outer map is faded out. Also, labels appear only within the masked feature. Now we can deactivate the zipcode layer, as we don't want its borders to be shown on the resulting map, and proceed to print composer to generate an atlas.

6. Open a new print composer and add to it a new map item and any other items you want to include in the layout. In our example, we will use the following:

   ° Custom page size equal to 200 x 200 mm.

   ° Road network map item with the same size, 200 x 200 mm, and a scale of 1: 20000. Don't lock the layers for this map item because it will be changed dynamically in the atlas.

   ° A small 50 x 50 mm map item in the top-right corner, with the New York borough boundaries layer and an overview added to it. After creating this map, lock its layers and their styles.

   ° Now you can come back to the main window and activate the layers that are necessary for the road network map.

7. After you have adjusted all the necessary items, go to the **Atlas generation** tab and activate the **Generate an atlas** toggle.

8. The **Configuration** section provides the following options to configure an atlas from its coverage:

   ° Select zipcode from the **Coverage layer** drop-down list. You can select any layer from this list, and its features will be used to iterate over them and create atlas pages.

   ° When **Hidden coverage layer** is active, the coverage layer isn't shown on the map. In our case, we don't use it because we have already deactivated the coverage layer.

   ° In the **Filter with** area, you can create an expression to select only some futures from the coverage area. For example, you can use only polygons with an area larger than some predefined value.

9. The **Output** section provides control over the results through the following options:

   ° **Output filename expression** is a template that will be used to name multiple output files.

   ° **Single file export when possible** generates a single multipage file (useful for exports to .pdf). If this option is active, **Output filename expression** is not applied.

- ◦ The **Sort by** option provides access to the names of attribute fields of the coverage layer, and you can select any of them to sort output files in ascending or descending order of the field values, with the arrow button beside.

10. Go back to the **Item properties** tab. Note that the map item with the main map (road network) should be selected. Activate the **Controlled by atlas** section, where you can select from among the following options:

    - ◦ **Margin around feature**: you can set the amount of surrounding area for each feature that is mapped on a single atlas page. This means that the map extent and scale will not be fixed. They will be adjusted to encompass the feature and its margin set. We select this option and set a value of 5 percent.

    - ◦ **Predefined scale (best fit)**: One of the project's predefined scales that best fits the feature into the atlas page will be used.

    - ◦ **Fixed scale**: Each feature will be shown in the same scale and the map page will automatically be centered around it. This is very useful when you need to print several maps in the same scale.

11. Now, everything is ready for you to generate and preview the atlas. You will notice that the **Preview atlas** button in the **Atlas** toolbar is active. Click on it to generate your atlas. When the generation process completes, you will see that the other buttons on the toolbar have also become active. Click on them to preview the atlas pages.

Using the toolbar buttons, you can not only explore your atlas, which contains multiple pages created within minutes, but also get quick access to the **Print** 🖶, **Export**, and **Settings** 🔍 options. The **Export** options are shown in this screenshot:

In the following screenshot, you can see an example of a single atlas page:

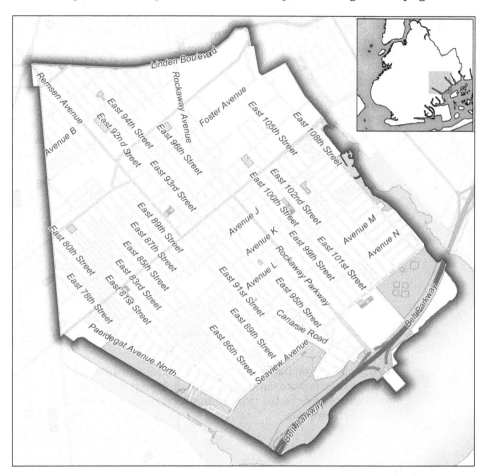

Managing print composers

Beginners usually forget to apply map locking options and often create every map from scratch. As a result, they are overloaded with multiple map composers that have permanently changing maps. If you want to save your time, then it is wise to use **Lock layers for map item** and **Lock layer styles for map item** to block undesirable changes in your maps.

You can have multiple maps and atlases that are stored within a single project, and get access to them by going to **Project | Print Composers**. Advanced options are available at **Print Composer Manager**, which is under **Project**.

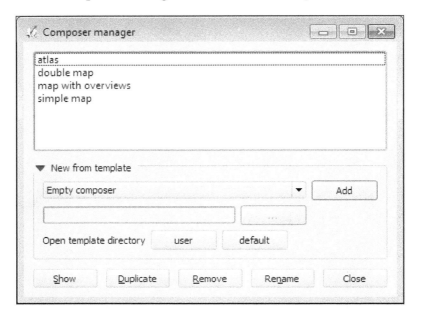

In **Composer manager**, you can create a new print composer from the template. Any map layout can be saved as a template from the print composer menu **Composer | Save as Template**.

Summary

In this chapter, you learned about basic map items and their configuration with print composer. Now you know that QGIS is not only a great tool for creating a simple print map, but it also provides sophisticated options to combine multiple maps into a single layout, relate them with overviews, and finally, generate multipage map atlases with a few clicks.

While print maps are very useful, they are not very efficient if you prefer to share your results online. In the next chapter, you will learn how to publish your data online quickly and efficiently.

Publishing the Map Online

4

In this chapter, we will cover the QGIS Cloud service, a powerful map hosting platform, and you will learn how to use it to share your maps with others by publishing them on the Internet. We will cover all the necessary steps, such as creating a QGIS Cloud account, preparing your maps for publishing, filling metadata, restricting access to information, and publishing the map. Also, you will learn how to use this published map from a web browser, QGIS, and other desktop GISes.

In this chapter, we will go through the following topics:

- Registering for the QGIS Cloud service
- The QGIS Cloud plugin
- Creating a database
- Publishing the map
- Viewing your map in QGIS and a browser
- Deleting unused maps

Registering for the QGIS Cloud service

QGIS Cloud is a cloud hosting service, developed and maintained by Sourcepole AG. Using this service, we can easily publish maps, share them with others, and even modify data—all this without any special knowledge, such as server setup and administration.

QGIS Cloud is built using open source technologies. To store vector data, it uses PostgreSQL/PostGIS databases, maps rendered with the QGIS server, and a web viewer based on QGIS Web Client. As a result, it integrates very well with desktop QGIS. Layers stored on the cloud can be loaded and edited with QGIS. A published map will have the same look and feel as a local map created in QGIS.

First, we need to create an account. This can be done with these simple steps:

1. Open your browser and go to `https://qgiscloud.com`.

2. Click on the **Sign Up** button on the main page.

3. Fill in and submit the registration form. You will need to provide a valid e-mail address and password.

4. After submitting the form, you will get a confirmation e-mail, with confirmation instructions. Confirm your account by visiting the link given in the e-mail.

That's all! Now you have a QGIS Cloud account with a free plan. This plan allows you to publish an unlimited number of public maps, create five PostGIS databases, and upload up to 50 MB of data.

 If you need more space for data, want to restrict access to your maps, or need some other special features, consider switching to one of the paid plans available. These plans, and the options they provide, are listed at `https://qgiscloud.com/en/pages/plans`.

You can also open the QGIS Cloud sign-up page directly from the QGIS Cloud plugin, if it is already installed. Just activate the plugin and click on the **Signup** link under the **Account** or **Services** tab. The corresponding page will open in your web browser.

The QGIS Cloud plugin

The **QGIS Cloud plugin** provides a user-friendly interface for publishing your data and maps with QGIS Cloud hosting. As it is a third-party plugin, we first need to install it.

If you have a proxy server, it is necessary to configure it first. **Plugin Manager** uses the same proxy settings as QGIS, so all you need to do is to go to **Settings | Options** and set up a proxy in the **Network** tab:

1. Start QGIS and open **Plugin Manager** by going to **Plugins | Manage and Install Plugins**. Wait while QGIS receives a list of plugins from the repository.

2. Switch to the **All** tab and start typing `cloud` in the **Search** field in order to filter the list of plugins, as shown in this screenshot:

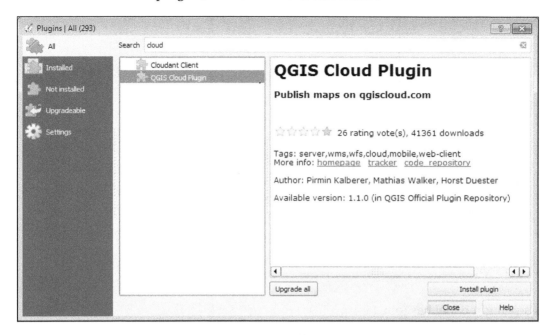

3. Then, select the plugin and click on the **Install plugin** button. The download will take some time, so be patient. When it completes, you will see a message bar with the confirmation of the download, and you can now close **Plugin Manager**. More information about installing plugins can be found in the *Extending functionality through plugins* section of *Chapter 1*, *Handling Your Data*.

After installing, the QGIS Cloud plugin will be available in the **Plugins** menu. QGIS Cloud plugin also puts button 🔘 in the **Plugins** toolbar and creates a **QGIS Cloud** floating panel. By default, this panel is docked on the left side of QGIS's main window.

Let's look at the plugin's interface:

The plugin panel has four tabs:

- **Services**: This is used to publish maps and update already published maps. Initially, this tab contains a short description of the plugin and the service itself, with links to the hosting website and its sign-up page. After publishing the map, you will find here links to the just published map (both desktop and mobile viewers) and the URL of the WMS service with your map.

- **Upload Data**: This allows you to upload local data to the QGIS Cloud database and publish the map. Here, we can select the database to be used, adjust the table names before publishing, and upload our local data to QGIS Cloud.

- **Account**: This is used for authentication and account management. With this tab, you can register for QGIS Cloud if you have no account yet. Also we can create databases and delete them here, if necessary.

- **About**: Here, you will find information about the plugin's authors, QGIS Cloud support contacts, and the plugin version. This information will be useful in the event of problems.

Now, switch to the **Account** tab of the **QGIS Cloud** panel, click on **Login**, and use your user name and password to log in. This may take some time, as the plugin needs to interact with the QGIS Cloud server. After a successful login, a message bar pops up, and you will see in the plugin panel information about your current plan, used/free disk space (if you already have databases), and a list of existing databases, if any.

Creating a database

QGIS Cloud stores all your geodata in PostgreSQL/PostGIS databases, so it is necessary to create at least one database for your data before uploading any layers and publishing the map.

To create a new database, perform the following steps:

1. Start QGIS and activate the QGIS Cloud plugin, if it is not yet activated.
2. Switch to the **Account** tab in the **QGIS Cloud** panel and log in using your username and password.
3. Click on the **Create database** button at the bottom of the **Account** tab and wait until the operation completes.

Database creation is a time-consuming operation, so be patient. When it's done, you will see a new database with a random name in the list of available databases. Also, the plugin registered this database in QGIS, so it can be used like any other PostGIS database, for example, for loading and editing layers from it in QGIS.

 If you are behind a firewall and/or have problems with creating the database, contact QGIS Cloud support via e-mail: support@qgiscloud.com.

In the event that this PostGIS connection — which is created by the plugin — is removed from QGIS by accident, it can be restored manually. Just log in to QGIS Cloud using the plugin and hold the cursor over the database name. In the tooltip, you will find all of the information (host, port number, database username, and password) needed for creating the connection. We can also use this information to connect to the database with other PostgreSQL clients, such as psql or pgAdmin.

Unfortunately, you cannot rename the database to something meaningful, so keep a list somewhere with information about which data is stored in each database, for future reference. Also, you can examine the database's content with the DB Manager.

For convenience, we recommend that you keep each map/project in separate databases. This introduces some disk usage overhead, as each empty database is about 11 MB in size, but such an approach also simplifies things a lot. For example, you can easily remove an unused map without losing data used in another map that is also stored in the same database. Of course, it is possible to keep many layers in a single database and use them in different maps, but this will lead to some complexity when you decide to remove some maps or layers.

Remember that you cannot create more databases than what is allowed by your plan; the same applies to using disk storage. If you need more space or databases, upgrade to another plan.

Publishing the map

Now, when a new database is created, we can publish the map designed in *Chapter 2, Visualizing and Styling the Data*, on the Internet.

Let's look at our project first; it contains raster and vector layers. There are several composers as well. Before publishing the data and map, we need to decide which information should be published and what we want to keep hidden.

By default, all vector layers are published with all their attributes available for both WMS and WFS services. All composers available in the project will be published too.

If you need to keep some attributes of the layer (or layers) unpublished, it is necessary to adjust the properties of the corresponding layer (or layers).

 Later, in the *WMS settings* section of this chapter, you will learn how to exclude entire layers and composers from publishing.

To do this, select a layer from the QGIS layer tree, right-click to open the context menu, and select **Properties**. In the **Layer Properties** dialog, go to the **Fields** tab, as shown in the following screenshot:

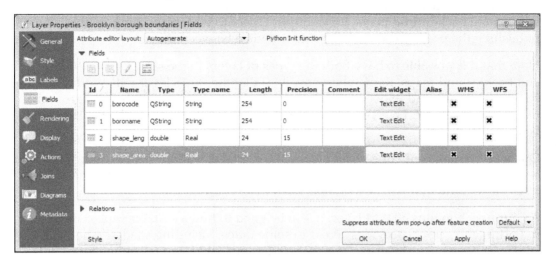

Here, we see a table with all the layer attributes, including additional information about them (data type, field length, precision, and so on). The last two columns, **WMS** and **WFS**, control attributes' visibility when the layer is published as WMS and/or WFS layer. If you don't want to publish some attributes, just deactivate the corresponding checkboxes. Then, click on the **OK** button to save the changes and close the dialog. Note that the same attribute can be published with WMS and not be available via WFS, and vice versa. Go through all the vector layers, examine their attributes, and hide them if necessary.

Now it is time to think about raster layers. Unfortunately, QGIS Cloud currently allows you to upload only local vector layers; local raster layer support is not available yet. But don't be discouraged! You can use WMS services and basemaps instead.

If our raster layers are also available as a WMS service, we can safely replace the local files with WMS layers, and our map will be published without losing any data. The already added WMS layers will be published as well.

But what to do if the raster data is available only as local files and you really need a background map? Fortunately, QGIS allows you to add basemaps; all you need is to install the **OpenLayers plugin**. Once this plugin is installed, go to **Web | OpenLayers plugin** and add the desired background maps from the available list. It is possible to add several background maps, for example, OpenStreetMap and Bing, so that after publishing the project, users will be able to switch between them.

Note that it is possible to have both the types of layers in the same map—WMS and background maps. Remember, however, that the WMS layer will overlay the background map and may cover it completely. If you want to see the background map under the WMS layer, don't forget to set the transparency for the WMS layers before publishing.

Our map contains only one raster layer—the digital elevation model—which has been loaded from a local file and, as a result, cannot be published. Let's add some other raster data that will be available from the published map.

As our map contains a DEM layer, it will be good to find a suitable replacement for it. Fortunately, USGS provides access to some maps, including shaded relief, via WMS. Of course, this layer is not as detailed as our DEM, but for online maps, it is more than enough. First, we need to create a new WMS connection by following these steps:

1. Click on the **Add WMS/WMTS Layer** button in the **Layers** toolbar, or go to **Layer | Add Layer | Add WMS/WMTS Layer**. In the **Add Layer(s) from a WM(T)S Server** dialog, click on **New** to create a new connection. Enter `USGS Relief (base)` as **Name** and `http://basemap.nationalmap.gov/arcgis/services/USGSShadedReliefOnly/MapServer/WMSServer?request=GetCapabilities&service=WMS` as **URL**, and click on **OK**. Select the newly added connection from the combobox and click on the **Connect** button.

2. From the layers tree, select the first node, called **USGS Shaded Relief...**.

3. In the **Coordinate Reference Systems** group, click on **Change**, find and select **WGS 84 / Pseudo Mercator (EPSG:3857)**, and click on **OK**.

4. Click on **Add** and close the WMS dialog.

5. Finally, move the added layer below all the vector layers, so that it will be at the bottom of the QGIS layer tree.

Note that this shaded relief layer contains data for a scale of 1:18000 or smaller. If you want to display relief on more detailed scales, it will be necessary to add one more layer using another WMS URL: `http://services.nationalmap.gov/arcgis/services/USGSShadedReliefLarge/MapServer/WMSServer?request=GetCapabilities&service=WMS`.

You may have noticed that we added the layer in **EPSG:3857** while all our vector layers are in EPSG:2263. As a result, we now have "on-the-fly" reprojection enabled, and the project CRS is EPSG:3857. This is because we will also add a background Bing map, which is available only in EPSG:3857. If you don't need a background map, it is better to use a CRS that matches the CRS of your vector data or a well-known CRS, such as **WGS 84** and **WGS 84 / Pseudo Mercator**.

Now, we will add a background map with Bing aerial images.

Note that using some background maps may violate the terms of service of those maps. Before adding any background map, check whether you are allowed to use that service with your online map and redistribute maps that contain data from that service.

To do this, perform the following steps:

1. Install and activate the **OpenLayers plugin** (see the *Extending functionality through plugins* section of *Chapter 1, Handling Your Data*, for more details).

2. Go to **Web | OpenLayers plugin | Bing Maps | Bing Aerial**.

3. Move the newly added layer to the bottom of the QGIS layer tree, so that it will be below all other layers.

If necessary, you can add more WMS layers and/or background maps, but don't overload your map by additional layers without the need to do so. Also choose which raster will be visible by default and turn the other raster layers off in order to speed up map loading. Save the project when all the changes have been implemented.

Now that our data is ready for publishing, our next step is project preparation. Of course, you can publish a map as is, but it is better to spend some time and make your map more user friendly and professional looking, especially if you plan to share it with others.

General settings

First, we fill in some metadata that will help users find our map and provide some information about it. To do this, open the **Project Properties** dialog by pressing *Ctrl + Shift + P*, or go to **Project | Project Properties...** and switch to the **OWS Server** tab. This tab allows us to configure the way in which our project will be processed by QGIS Server. There are many different settings here; right now, we will look at the **Service capabilities** group:

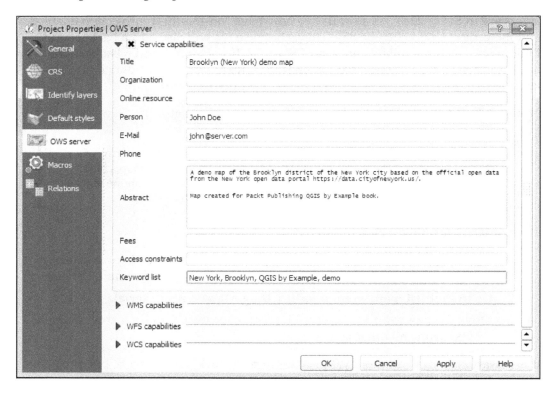

Here, we need to fill in this metadata:

- **Title**: This is the title of the map. We select Brooklyn (New York) demo map.

- **Organization**: Here, we specify the map author or owner. Enter your name here.

- **Online resource**: Here, we specify the URL that should be used to access our map. Right now, we can leave it empty because the map is not published, which is why we don't know its URL. But if you are curious, here is the convention used for URL generation in QGIS Cloud: `http://qgiscloud.com/<user_name>/<map_name>/wms`. In this, `<user_name>` is your QGIS Cloud login and `<map_name>` is the filename of the published project without extension. So, suppose your login is alex and the project will be saved after uploading the data as `brooklyn_demo.qgs`. Then, the URL will be `http://qgiscloud.com/alex/brooklyn_demo/wms`.

- **Person**: This is the name of the person who is responsible for the published map. It is mainly used as support contact. In our case, this is the same person as specified in **Organization** field.

- **E-Mail** and **Phone**: These are the contact details of the responsible person. We can leave them empty or fill in only one field so that users will be able to contact the creator of the map if they have any questions or problems.

- **Abstract**: This is a brief description of our map and used data. It is important information, as it helps users understand the purpose and coverage of the map. Such information can also be used by metadata catalogs for the purpose of indexing and searching maps. Here is our description; you can use it as is or modify it according to your tastes:

A demo map of the Brooklyn district of the New York city based on the official open data from the New York open data portal `https://data.cityofnewyork.us/`.

Map created for Packt Publishing *QGIS by Example* book.

- **Fees**: This shows information about the fees. It can be ignored if map does not use fees.

- **Access constraints**: This describes the restrictions on and legal requirements for using this map. For example, it can be "For internal use only" or some copyright notice. For public maps, this field usually is empty.

- **Keyword list**: This is a list of keywords or short phrases that describe the map. This information helps in searching and indexing. For our map, we use these keywords; feel free to adjust them to suit your tastes:

`New York, Brooklyn, QGIS by Example, demo.`

It is good practice to fill in metadata not only for the map but also for each layer in it, so that when the user looks for the available WMS or WFS layers, they will see the layers' descriptions and easily understand that a particular layer is suitable for them. Anyone will agree that working with a WMS server where layers have a title and an abstract is more comfortable than working with a server without such information.

To edit the layer's metadata, select it from the QGIS layer tree, right-click to open the context menu, and choose **Properties**. In the **layer properties** dialog, go to the **Metadata** tab. The most important pieces of metadata to fill in are **Title**, **Abstract**, and **Keyword list**. After editing the metadata, click on **OK** to close the dialog and save the project.

WMS settings

Now look at the **WMS Capabilities** group, which allows you to tune the WMS service that will be created from our map, as shown in this screenshot:

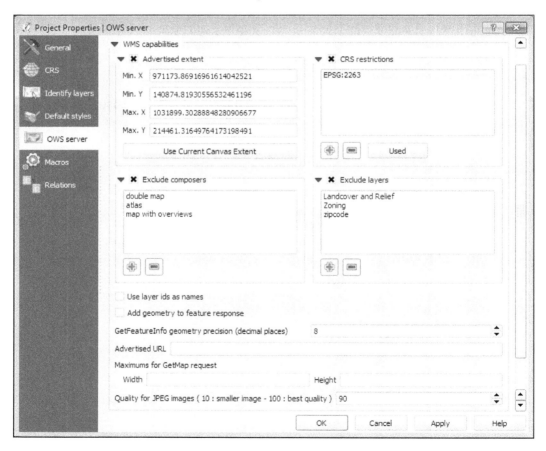

By default, the entire map will be published. If you want to make only a part of your map available via WMS, check the **Advertised extent** box and define the allowed extent by entering its minimum and maximum coordinates. Click on the **Use Current Canvas Extent** button to restrict the available extent to the current map extent.

QGIS Server, used by QGIS Cloud, implements a special extension to the WMS standard: the **GetPrint** request. This allows you to create printed maps using one of the composers available in the project. By default, all available composers will be published and advertised for printing. If you don't need this functionality at all or want to hide some composers from users, activate the **Exclude composers** group and add the composers that need to be excluded to this list using the **+** button.

Similarly, you can hide some layers if you don't want to publish all of them. To do this, activate the **Exclude layers** group and add layers to exclude using the **+** button.

> Of course, we can just remove the layers, save the project with a different name, and then publish this new project. But in this case, we will lose one major advantage—the ability to use the same project for offline and online maps.
>
> Imagine that after publishing a map, you need to change the symbology of a layer. In the case of two different projects, you will need to keep them in sync. It is much simpler to have a single project and hide the layers by changing the settings.

The next important setting is **CRS restrictions**. By default, every layer can be loaded in any QGIS-supported **coordinate reference systems (CRS)**. As a result, the service capabilities document becomes very large, especially when the map contains many layers. This in turn causes slow map loading in web-based viewer and desktop clients, as they have to download and parse much more data.

In real-life applications, only a small subset of coordinate reference systems is required. So, we can safely restrict the list of available CRSes without any drawbacks, and moreover speed up the loading of our map. To do this, activate **CRS restrictions** and add all the necessary CRSes. The **Used** button allows us to add the current project's CRS to the list in just one click.

If you want to enable feature highlighting on identification, don't forget to select the **Add geometry to feature response** checkbox. It is also possible to limit coordinate precision by adjusting its value in the **GetFeatureInfo geometry precision (decimal places)** field.

Finally, you can limit the size of the image returned by the GetMap request by defining the width and height in the **Maximums for GetMap** request and adjusting **Quality for JPEG images**.

WFS settings

If you want to make your data available not only as rasters but also as vectors, it is necessary to configure **WFS capabilities**, which are shown in the following screenshot:

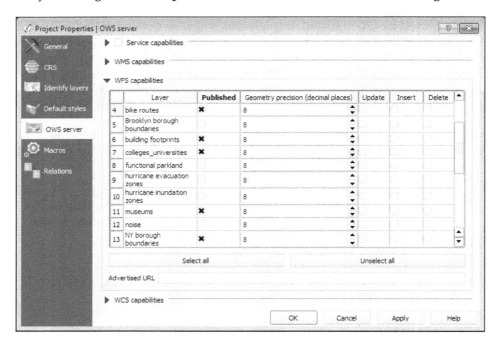

All you need to do is select the layers that should be available via WFS by selecting the corresponding checkboxes in the **Published** column. If the layer needs to be editable, it is necessary to specify the supported operations by activating the checkboxes in the corresponding columns:

- **Update**: The user can edit existing features (both geometry and attributes) in layers with this capability

- **Insert**: The user can add new features to layers with this capability

- **Delete**: This allows the user to delete existing features from a layer

Remember that with the QGIS Cloud free plan, your map will be publicly available, and anyone will be able to edit your layers. If you need to restrict access, either consider switching to other plans or don't enable editing via WFS.

Being the map owner, you can always edit published vector layers by connecting to the corresponding QGIS Cloud database via QGIS.

When you are satisfied with all the settings, click on the **Apply** button to save your changes and then close the **Project Properties** dialog.

Now, we can publish our map. Activate the QGIS Cloud plugin, if it has not already been activated. Switch to the **Account** tab in the **QGIS Cloud** panel and log in using your username and password. First, we need to create the database where all our layers will be saved. Database creation is covered in detail in the *Creating a database* section of this chapter; return to it if necessary. We will assume that a new empty database already exists.

Switch to the **Upload data** tab. From the **Database** combobox, select the database created previously. In the table below this combobox, you will see all the layers supported by QGIS Cloud that can be uploaded. If there are no layers, click on **Refresh layers**. If necessary, you can adjust the name of the database table used to store layer data by editing the text in the **Table name** column.

 Note that while all other columns are editable too, you should not edit them.

Also verify that all the layers listed in this table have the correct CRS assigned. QGIS Cloud supports the CRSes available in the EPSG database only, so if some layers have a user-defined CRS, it is necessary to reproject them before publishing.

If **Replace local layers in project** is checked, the plugin will update the project by replacing all local vector layers with uploaded layers. So, your project will contain layers from the QGIS Cloud database, and you can easily edit them from QGIS. All changes in these layers will be immediately visible in the published map.

To start the data upload, click on the **Upload data** button. The upload will take some time, depending on the data size and speed of your Internet connection. When it completes, you will be prompted to save the updated project. We recommend that you save the published project with a different name so that you will have the project with the local data unchanged.

 QGIS Cloud uses the name of the project file as the map name, so choose the filename of your project carefully; it's better to use something meaningful.

After successful data upload, a new **Publish Map** button will be shown. Click on it to publish your map. That's all! Now your map is available on the Internet. Switch to the **Services** tab to get the URLs of standard and mobile viewers, the WMS service (the same link can be used for WFS), and the web interface admin panel.

If necessary, we can easily update the published map; for example, we can change layer symbology, or hide or show some attributes or composers, and so on. To do this, just open your project containing the Cloud-based layers, make necessary changes, and then publish the map by following these steps:

1. Activate the QGIS Cloud plugin, if it is not activated yet.
2. Switch to the **Account** tab in the **QGIS Cloud** panel and log in using your username and password.
3. Now switch to the **Services** tab and press the **Publish Map** button to publish the updated map.

There is no need to upload the data again, as all of the data was uploaded previously, and our project now contains layers from the QGIS Cloud database instead of local files.

Viewing your map in QGIS and a browser

A published map can be viewed in QGIS and a browser. Depending on your needs, it is better to use one or the other. A web-based viewer is a good choice if all you need is viewing and some basic GIS functionality, such as map navigation, feature identification, and printing. Also, a web-based viewer is a great way to share your map with others.

Working with a map from a browser

QGIS Cloud provides two web viewers: one for mobile devices (smartphones or tablets), and another for desktop. Both viewers provide the same basic functionality such as map navigation and changing layer visibility. Let's take a closer look at them.

The standard desktop viewer is based on QGIS Web Client and it looks like a simple GIS application that runs in the browser, as seen in the following screenshot:

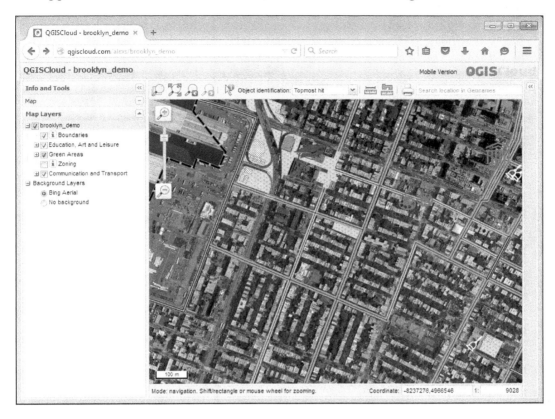

There is a layer tree on the left side, where you can change the layer's visibility and enable or disable background maps, if any. On the right-hand side, we can find the **Attribute Data** panel (hidden by default), which is used to display information about the identified feature. The rest of the space is occupied by the map view itself.

In the preceding screenshot, you can see a toolbar with navigation buttons, the identify tool, tools for measuring length and/or area, and the **Print Map** button. The navigation, identify, and measuring tools work similar to the corresponding tools from QGIS.

The most interesting feature of the standard viewer is map printing. If published project contains composers, they can be used for producing printed maps. They can be used to produce printed maps. To create such a map, zoom to the area of interest using the navigation tools, and then click on **Print Map** button. A **Print Settings** dialog will appear, and a selection rectangle will be shown on the map, like this:

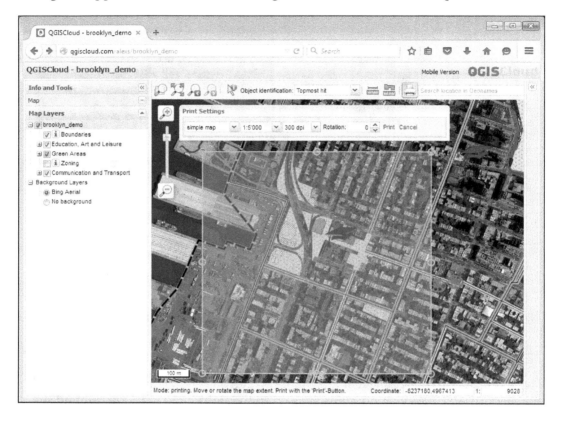

Define the boundaries of the region you want to print by moving and resizing the rectangle. Then use the floating **Print Settings** panel to select the desired map layout, scale, and resolution. Define the map rotation, if necessary. When all the settings are made, click on the **Print** button to generate a PDF file containing your map. The generated map will be opened in a new browser window, and can be saved on the disk for further usage.

The mobile viewer, as the name implies, is optimized for usage with mobile devices, such as tablets or smartphones.

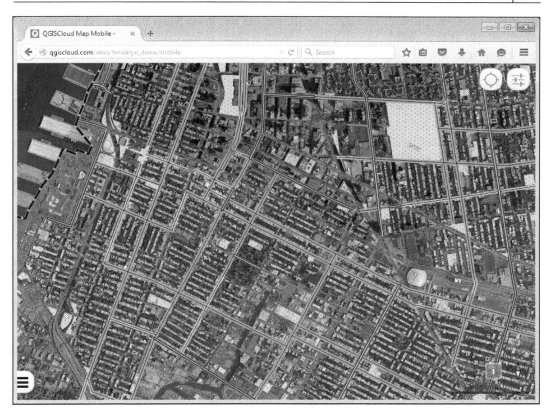

All of the available screen area is used to display the map. The buttons on touchscreens are large and easy to use. To access the layer tree, press the button in the bottom-left corner. The buttons in the top-right corner allow you to enable location tracking and set up some other options such as map rotation, scalebar display and so on. With the mobile viewer, you have no access to the print map capability. Nonetheless, it is possible to use the map for navigation. The viewer will track your position and update the map when it changes. If necessary, it is also possible to activate map rotation so that the map can be rotated to match its north to the actual north.

Working with a map in QGIS

Web-based viewers are good if all you need is read-only access to data. A published map, when used from QGIS, gives you more functionality and flexibility.

 You can use published maps not only from QGIS but also from any WMS or WFS client.

By default, every published map is available via the **Web Map Service (WMS)** protocol. So, it is possible to use a layer from it as a background layer in another map or embed it in a web page with the help of libraries such as OpenLayers and Leaflet.

To add a layer from a published map as a WMS layer to QGIS, perform the following steps:

1. Open the **Add Layer(s) from a WM(T)S Server** dialog using the corresponding button on the **Manage Layers** toolbar, or from the **Layer** menu (**Layer | Add Layer | Add WMS/WMTS Layer...**). Add a **New** connection.

2. Enter a connection name (it's best to use something meaningful) and the URL of the map.

3. Select the created connection from the list and click on **Connect**. A list of available published layers will appear. If the map layers have metadata filled in, you will see their titles and descriptions.

4. Select the necessary layers and add them to your project.

Note that you can add only layers created from your local data. Background maps, and WMS layers used in published map are not available. If you need a background map and/or WMS layers, add them by using the corresponding plugins or the QGIS core functionality.

Also, if enabled by the author, a map can be accessed via the **Web Feature Service (WFS)** and even modified with the **Web Feature Service Transactional (WFS-T)** protocols. To access a map via WFS-T with QGIS, perform the following steps:

1. Open the **Add WFS Layer from a Server** dialog using the corresponding button in the **Manage Layers** toolbar, or from the **Layer** menu (**Layer | Add Layer | Add WFS Layer...**).

2. Add a **New** connection.

3. Enter a connection name (it's best to use something meaningful) and the URL of the map.

4. Select the created connection from the list and click on **Connect**. A list of available published layers will appear.

5. Select the necessary layers and add them to your project.

If the selected layers have update, delete, and insert capabilities, you can edit them with QGIS just as would do with any other layer. All changes will be visible in the published map immediately.

Deleting unused maps

With the passage of time, some maps become unnecessary for various reasons. If you don't need some maps and data, you will be pleased to know that it is possible to remove them and, as a result, free some disk space for new projects.

There are several ways of removing unused data: with the QGIS Cloud plugin, from the QGIS Cloud web interface, and by using DB Manager or any other PostgreSQL/PostGIS client. They work differently and complement each other.

Deleting layers with DB Manager

As we mentioned previously, QGIS Cloud uses PostgreSQL/PostGIS to store spatial data. So, we can easily connect to the database with **DB Manager** or any other PostgreSQL client (for example, pgAdmin) and delete unused tables.

Assuming that a connection to the QGIS Cloud database is already created, here are the necessary steps that we need to perform in order to delete layers, using the **DB Manager** plugin:

1. Start the **DB Manager** plugin. As this is a core plugin and is activated by default, you can find it by going to **Database | DB Manager**. If the **Database** menu does not exist, make sure that the **DB Manager** plugin is enabled. More information about using QGIS plugins can be found in the *Extending functionality through plugins* section of *Chapter 1, Handling Your Data*.

2. Expand the **PostGIS** item in the database tree and find your database.

3. Expand the **public** schema and find the table you want to delete.

4. Select the table and right-click to open the context menu. Select **Remove** to delete this table from database.

Note that you also need to delete or update the corresponding maps after deleting tables. See the *Deleting maps from the web interface* section later in this chapter for more information on this.

Deleting databases from the QGIS Cloud plugin

The QGIS Cloud plugin allows you to not only create a database and publish data and maps, but also remove unused databases. To remove an existing database, simply follow these steps:

1. Start QGIS and activate the QGIS Cloud plugin, if it is not activated.

2. Switch to the **Account** tab in the **QGIS Cloud** panel, and log in using your username and password.

3. In the **Databases** list, find the database you want to remove and select it. Click on the **Delete database** button and wait while the operation completes.

 Be careful! Currently, the QGIS Cloud plugin allows you to remove only an entire database, so you will lose all of the data from other maps that are stored in the deleted database. Make sure you have backups, or look at the previous section to learn how to delete separate layers.

Note that the plugin deletes the entire database, with all of the data in it. Maps created using data from this deleted database will still be listed in your account, but they will be empty. To delete these maps, we need to use the QGIS Cloud web interface, as described in the next section.

Deleting maps from the web interface

Using the web interface, we can remove only unused maps. All of the data associated with them will be kept. So, this is a supplementary method and it should be used with the methods already described.

To remove an unused map, follow these steps:

1. Open the browser and go to the QGIS Cloud site at `https://qgiscloud.com`.

2. Log in to your account using your username and password.

3. Click on the **Maps** link in the menu at the top.

4. Locate the map that should be removed and click on **Destroy**. Confirm your action in the confirmation dialog.

Remember that you've deleted only the map. The data used in this map is still available in the database (provided that you haven't deleted the database first). So, if you don't need this data, it is necessary to delete the database too, unless it is being used by other maps. If you are using the same database to store data for different maps, you can delete separate layers with any PostgreSQL client, as described in the *Deleting layers with DB Manager* section.

Summary

In this chapter, we covered how to publish our maps online using QGIS Cloud hosting from QGIS. We saw how to prepare projects for publishing, including specifying project metadata, restricting access to some layers and composers, hiding layer attributes, and defining allowed actions. Also, you learned how to access a published map from different programs and how to manage your QGIS Cloud account.

In the next chapter, you will start getting familiarized with the analysis capabilities of QGIS. The first analysis type you will get to know is density analysis.

5
Answering Questions with Density Analysis

We often need to work with large and dense datasets in which there is a lot of overplotting and features experience significant overlap. Such datasets may be very slow at rendering because they contain thousands, or even millions, of features, and very difficult to interpret because overlaps make it difficult to detect any clusters or distribution patterns. In this chapter, you will learn the techniques that allow you to visualize such datasets in a more readable and faster way by displaying feature density instead of the features themselves. By the end of this chapter, you will be able to perform density analysis of your data and extract information from density maps.

In this chapter, we will go through the following topics:

- Density analysis and heat maps
- Creating heat maps with the Heatmap plugin
- Mapping density with a hexagonal grid

Density analysis and heatmaps

Density maps allow visual estimation of object or event concentration over the study area. Such maps are very useful for assessment of the distribution patterns of the features over the study area. When we simply add locations of the features or events (for example, as points) to the map, we cannot see the changes in their concentration in different areas. Density analysis gives us such functionality by using uniform area characteristics, such as feature count per acre or square kilometer.

A density map gives us the ability to estimate the concentration of some features within an area. This helps us find areas where an urgent reaction is required or which match your criteria. Heatmaps also help control conditions and their changes.

Density maps are also extremely useful when mapped regions (for example, districts) have different sizes. For example, if we want to know how many people live in each district, we just need an ordinal map with the population data. According to this map, a large district may have a higher population than a smaller district. But if we want to identify the districts with a higher concentration of population, then we need a density map to see the number of people per square kilometer. And a density map will show us that, in fact, small regions with a high population density may have more people per square kilometer than larger districts.

Generally, we can show on a map the density distributions of the features themselves (for example, schools), as well as distributions of some numerical characteristics of these features (for example, the number of pupils in schools). The results will be completely different in these cases. A density map of schools can help an education department find areas where more schools are needed, while a density map created from information about the number of pupils in each school may help a transportation company to plan bus routes and to decide where to place bus stops.

The most common use case is the creation of density maps to display the density of point features. Such maps are often called **heat maps**. What is a heat map? It is a raster layer. Each cell of it contains a representation of the density of features in its vicinity (for example, the number of people per square kilometer), which depends on the number of features within some area.

To create a heat map, in the simplest case, GIS looks at the features around a cell center, using a given search radius. Then the number of features that fall within the given radius is calculated and divided by the area of the region. This value will be a cell value. Then next cell will be analyzed, and so on. As a result, we will get a combination of values, which creates a smooth surface. To understand this better, refer to the following diagram:

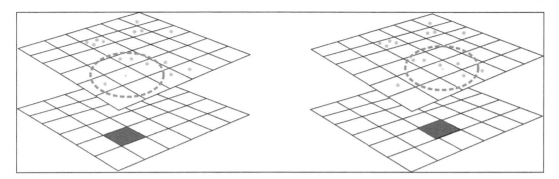

This diagram shows the general principle of creating heat maps. The green dots depict the features used for density map generation, the blue square is a current raster cell, and the red dotted circle marks the search radius, for example, 1 km. In this case, the area covered will be about 3.14 sq. km. As we can see in the diagram to the left, four features are within the search radius. So, the raster cell will get a value of $4/3.14 = 1.27$. On the right side, we notice that the next cell will get a value of 1.59 because now there are five features inside the search radius.

This is the simplest approach. In real-world applications, more complex algorithms are used, where each point has some impact on the values of the neighboring cells, depending on its distance from those cells.

Creating heat maps with the Heatmap plugin

With the help of the QGIS core plugin called Heatmap, we can easily create heat maps from vector point data and use it for further analysis. First, we need to activate this plugin, if it has not yet been activated. After activation, it creates a submenu under the **Raster** menu and places its button on the **Raster** toolbar.

Let's create a density map for the noise layer, which contains information about complaints regarding high noise levels. This layer contains 44,397 features, and it is difficult to know which places are noisy.

Information about such places may be useful for a police department or other agencies to plan some activities for noise reduction, or for those who are looking for apartments and don't want neighbors who like playing loud music:

1. Start the plugin by clicking on the **Heatmap** button on the **Raster** panel, or by navigating to **Raster | Heatmap | Heatmap...**.

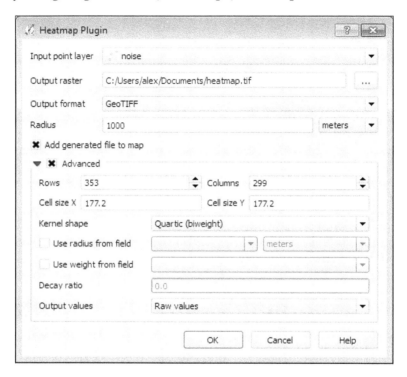

2. Select the noise layer from the **Input point layer** combobox.

3. Using the **...** button on the right side of the **Output raster** field, specify the location in which the resulting heat map needs to be saved. Note that there is no need to specify the file extension; it will be picked up automatically, based on the output file format.

4. Use the **Output format** combobox to select the desired format for the heat map. The most common choice here is **GeoTIFF**, but for very large maps, it is better to use something different, for example, **Erdas Imagine**.

5. The last thing we need to specify is **Radius**. This value defines the distance from each cell up to which QGIS will look for neighbor features and take their presence into account. Generally, a bigger search radius gives a more generalized result, as the number of features found will be divided by a bigger area. A smaller radius gives more precise results, but if this value is too small, we may not find any distribution patterns. The search radius can be defined in meters or map units.

To determine the search radius from a known area, we can use a very simple formula derived from formula for the area of a circle:

$$r = \sqrt{\frac{S}{\pi}}$$

For example, if we need to calculate density per square kilometer, then the search radius will be as follows:

$$r = \sqrt{\frac{1km^2}{\pi}} = \sqrt{\frac{1000000m^2}{3.1415926}} \approx 564.2m$$

For more fine-grained control over the result, we can check the **Advanced** box and define some additional parameters:

- **Rows** and **Columns**: These allow us to define dimensions of the output raster. Larger dimensions will result in a bigger output file size, while smaller dimensions will result in a rough and pixelated output. Input fields are linked to each other, so changing the value in the **Rows** field (for example, halving it) will also cause the corresponding change to the value in the **Columns** field, and vice versa. Furthermore, these values have a direct influence on the raster cell size (see the next point). It is worth mentioning that the extent of the raster preserved when changing raster dimensions.

- **Cell size X** and **Cell size Y**: The raster cell size determines how coarse or detailed the display of the distribution patterns will be. A smaller cell size will give smoother results, but the processing time and memory required for the analysis will increase. Large cells will be processed faster, but the resulting raster will be pixelated. If the cells are really big, some patterns will become invisible, so you may need to run the analysis several times, trying different cell sizes to get results that satisfy your requirements.

  The cell size depends on and is linked to the raster dimensions. Increasing it will decrease the number of rows and columns, and vice versa.

- **Kernel shape**: This controls how the point influence changes with changes in distance from this point. The QGIS Heatmap plugin currently supports the following kernels:

    ° quartic (also known as biweight)

    ° triangular

    ° uniform

    ° triweight

    ° Epanechnikov

The following diagram shows the distribution of the point influence for different kernels:

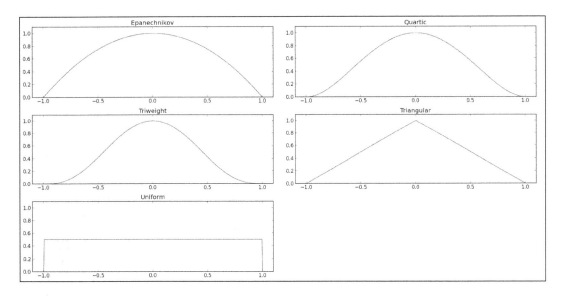

> For more details about kernel shapes, refer to the Wikipedia article at `http://en.wikipedia.org/wiki/Kernel_ (statistics)#Kernel_functions_in_common_use`, and books about statistics, for example, *Density Estimation for Statistics* and *Data Analysis* by B. W. Silverman.

Depending on the kernel shape, we will get a smoother heat map, or more clearly exposed hotspots. For example, the triweight kernel will give clearer, sharper hotspots than the Epanechnikov kernel, because the Epanechnikov kernel has lower influence near the hotspot center. Also, in different scientific fields, different kernels are preferred; for example, in crime analysis, the quartic kernel is typically used.

It is also possible to use a variable search radius for each point by selecting the **Use radius from field** checkbox and selecting the attribute field with radius value from the combobox. If you need to weight points (in other words, increase or decrease their influence) by some numeric attribute, activate the **Use weight from field** checkbox and select the corresponding field. In our example, we will not use this functionality, but you can try it yourself.

As we said before, cell size has a direct influence on the quality of the resulting heat map, so it is important to select it carefully. In most cases, the cell size is chosen in such a way that we get 10 to 100 cells per unit area (which in turn is defined by the search radius). To calculate the cell size, we need to align area units with distance units; for example, if we calculate the density using square kilometers and define the search radius in meters, then it is necessary to convert the square kilometers to square meters. The next step is to divide the area by the desired number of cells. Finally, as the cell size is defined by its width or height (because raster cells usually have a square shape), we need to extract the square root of this value.

In our example, we will create a heat map with a search radius of 1000 m, so the lookup area will be approximately 3.14 square kilometers. When expressed in meters, this will be as follows:

$$3.14km^2 = 3.14 \cdot 1000m \cdot 1000m = 3140000m^2$$

As we want a smooth heat map, we will use a relatively large number of cells per unit area; let's say 100 cells per 3.14 square kilometers. So, we divide the area in square meters by the desired cell count:

$$\frac{3140000m^2}{100\,cells} = 31400m^2\ per\ cell$$

Finally, we calculate the square root of this value to get the cell size that allows us to have 100 cells per 3.14 square kilometers:

$$\sqrt{31400m^2} \approx 177.2m$$

Of course this is not a strict rule but just a recommendation. You can safely use another cell size, depending on your data and the desired results. Just remember that smaller values lead to smoother heat maps, but at the same time increase the analysis time and output raster size.

When all the inputs and parameters are set, press the **OK** button to start the process of the heat map generation. The progress of the heat map formation will be displayed in a small progress dialog. If this process is taking too long time to complete, you can interrupt it by pressing the **Abort** button. Note that after aborting heat map generation, you still get the output, but it will be incomplete and not useful for further analysis.

When the process completes, the generated heat map will be added to QGIS as a grayscale raster, where lighter regions correspond to higher density values and darker regions correspond to lower density values, like this:

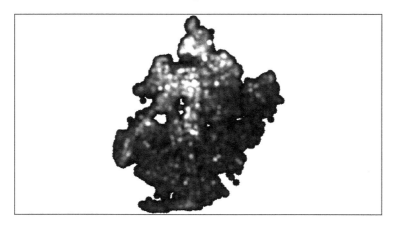

To improve readability and make it look like real heat map, we need to change its style. To do this, follow the next steps. For more detailed information about styling raster layers, refer to the *Developing styles for raster layers* section of *Chapter 2, Visualizing and Styling the Data*:

1. Right-click on the `heatmap` layer in the QGIS layer tree. In the context menu, select **Properties**.

2. Go to the **Style** tab and select **Singleband pseudocolor** as **Render type**.

3. In the **Load min/max values** group, activate the **Min/max** options. Set **Extent** to **Full** and **Accuracy** to **Actual (slower)**. Press the **Load** button to get raster statistics. This will be used for further classification.

4. Select a suitable color ramp in the **Generate new color map** group, for example, `YlOrBr` (which changes colors from yellow to orange and then brown), or `Reds` (which uses different shades of red). If necessary, change the number of classes and click on the **Classify** button.

5. Click on **OK** to apply the changes and close the properties dialog.

Now we can easily locate the hottest points (displayed in colors closer to red if the Reds color map is used), and even recognize some distribution patterns that were not visible when we looked at the original point layer:

Now we can easily locate the hottest points (displayed in colors closer to red if the Reds color map is used), and even recognize some distribution patterns that were not visible when we looked at the original point layer. Also, our heatmap layer showed up much faster than the vector which is used to create this heat map.

Detecting the "hottest" regions

Sometimes, you don't need the heat map itself, but just want to find the hotspots—areas with the highest density—and use them in further analysis. It is pretty easy to find such regions in QGIS and extract them in the vector form.

First, we should define threshold value, which will be used to recognize the hotspots. As a starting value, we can use the maximum pixel value in our heat map and then adjust it as per our needs.

The simplest way to find the maximum pixel value is to use the **Identify Features** tool. Select a layer in the QGIS layer tree, activate the **Identify Features** tool, click on the most visually "hottest" regions, and look at the reported value. With our heat map, this will be 540.32.

If we will use this value as is, we cannot find all the important clusters, so this value should be reduced first. The smaller the selected value (in comparison with the maximum value), the larger the number of clusters found. The area of separate clusters will also grow. For our example, we choose a value of 200.

Now, open **Raster Calculator** from the **Raster** menu, specify a path where the output file should be saved in the **Output layer** field, and enter the `"heatmap@1">=200` formula in the **Raster calculator expression** field, like this:

This formula is used to create a so-called mask. If the pixel value of the input layer is greater or equal to our threshold value of 200, then the output pixel value will be 1. Otherwise, it will be 0. So, our output raster will be a binary raster, with only two pixel values — 0 and 1 — which is very easy to convert to a vector.

Leave all other values unchanged so that the resulting raster will have exactly the same dimensions and cell size as the input one. Press the **OK** button to start the calculation. When it is done, a new black-and-white raster layer will be added to the QGIS canvas, as shown here:

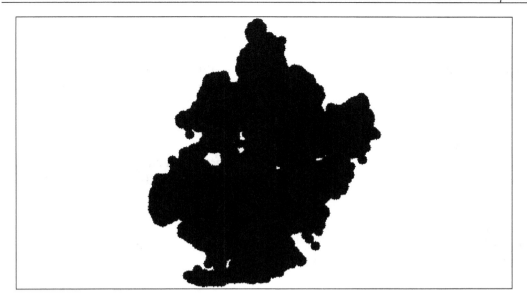

To convert the mask raster into vector format, we need to create polygons from all connected pixels with the same value. This is where the **Polygonize** tool comes to help. In the **Processing** toolbox, you can find the **Polygonize** algorithm by typing its name in the filter field at the top of the toolbox. Double-click on the algorithm name to open its dialog, and you will see something like this:

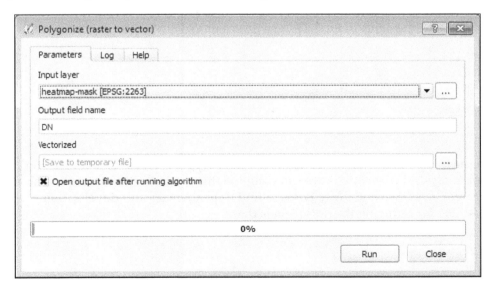

Select the mask layer that was previously created as **Input layer**, specify the path where the result will be saved using the **Output layer** field, and click on the **Run** button to start the algorithm. When it is done, a new vector layer will be added to QGIS. This layer has an attribute called DN (if you did not change it) that indicates the pixel value of each polygon in the layer. So, all we need to do is remove all the features that have the attribute value equal to zero. The remaining features will be the hotspots.

To delete unnecessary features from the hotspots layer, select it in the QGIS layer tree, right-click to open the context menu, and select **Open Attribute Table**. Click on the **Select features using an expression** button. In the **Select by expression** dialog, enter "DN" = 0 (if necessary, replace DN with your field name), click on the **Select** button, and close the dialog. Start editing by clicking on the **Toggle editing mode** button, or press *Ctrl* + *E*. To remove the selected features, press the *Delete* key or click on **Delete selected features**. Finally, turn editing mode off by pressing *Ctrl* + *E* or clicking on **Toggle editing mode** again.

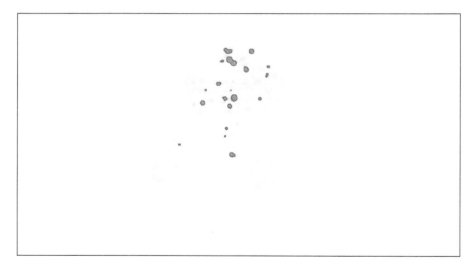

Now, the hotspots layer contains only hotspot polygons, which can be used for further analysis. For example, we can combine this cluster with information about the nearest buildings and noise types to find dependencies and develop some suggestions for reducing the noise levels there.

Looking for distribution patterns with contour lines

Apart from detecting hotspots, heat maps can also be used to detect intensity changes or visualize the direction of value changes. The most common way to do both of these tasks is contour lines generation.

Fortunately, QGIS has all the necessary tools for this. We will use processing again, but contour lines generation is also available in the **GDALTools** plugin (which can be found in the **Raster** menu). In the **Processing** toolbox, you can find the **Contour** algorithm by typing its name in the filter field at the top of the toolbox. Double-click on the algorithm name to open its dialog, which looks like this:

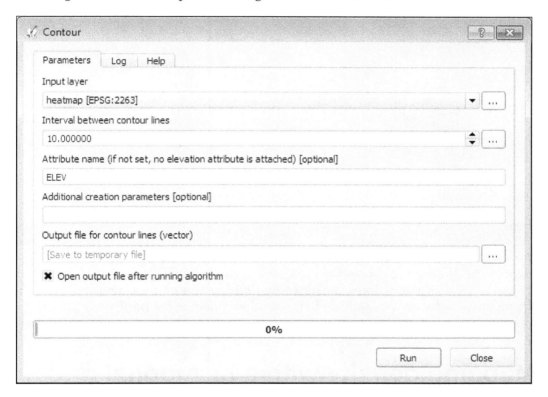

Select the heatmap raster layer as **Input layer**. In the **Output file** field, specify the path where the results will be saved. Also, it is necessary to define **Interval between contour lines**. There are no strict principles about determining this interval. The general rule is to select such an interval that detects patterns in the areas with smooth density changes. We will select an interval of 10.

When all of the necessary information has been defined, click on **Run** to start the contour line generation. After some time, a new polygonal vector layer will be added to QGIS, and we can start analyzing it. First, if necessary, move the contours layer to the top of the heat map in the QGIS layer tree. Also, it is better to adjust contours' symbology to make them more recognizable against the background of the heat map.

More dense contours correspond to more intense density changes. Also, we can identify the direction of changes in noise. For example, in the preceding screenshot, we can see that in some places, the noise distribution around the center is not equal; the intensity reduces more quickly in the southeast than in the northwest. So, we may assume that there are some obstacles to noise there.

Mapping density with a hexagonal grid

There is also another approach of mapping density, called binning. Generally speaking, binning is a technique of grouping N values/features into M groups, where $M < N$. The result of such an operation can be interpreted as a two-dimensional histogram.

Binning is an alternative to heat maps, not a replacement. The choice of the method depends on the requirements and further usage of the results. It is, however, worth mentioning that binning produces a vector output, while a heat map produces a raster output.

In general, binning can be described in two simple steps:

1. Create a hexagonal grid on top of the point layer.
2. Count the number of points in each grid cell.

In this section, you will learn how to use this technique in QGIS in the example of hexbinning, in other words, mapping density with a hexagonal grid.

 In fact, for binning, we can use not only hexagons but also other shapes that allow regular tessellation of a 2D surface – triangles and rectangles.

Why do we choose hexagons? Well, because a hexagon is closest to a circle among all shapes that tessellate. As a result, they represent curves more naturally. Another advantage of hexagons is a more compact structure, so the distance between cell centers in a hexagonal grid is lower than in a rectangular grid. So data aggregation around the cell center is more efficient.

 In *Chapter 8, Automating Analysis with Processing Models*, we will create a model that produces two density maps: hexagonal and rectangular, so that you can compare them side by side and better understand their differences and use cases.

In the upcoming sections, we will create a density map using this approach. For this exercise, we will use data from Brooklyn's street tree census. This is the trees layer in our map.

Creating a hexagonal grid

To create a hexagonal grid, we will use the QGIS Processing framework and its algorithm called **Create grid**:

1. In the **Processing** toolbox, find the **Create grid** algorithm by typing its name in the filter field at the top of the toolbox. Double-click on the algorithm name to open its dialog, which looks like this:

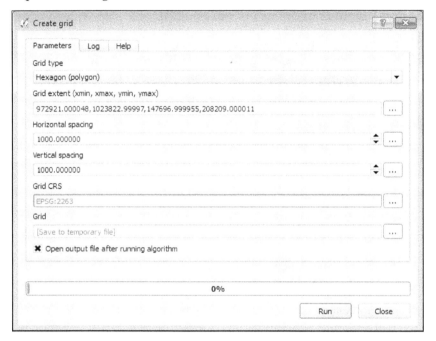

2. In the **Grid type** combobox, select **Hexagon (polygon)**.

3. To specify grid extent press the **...** button on the right side of the **Grid extent** field and choose **Use layer/canvas extent** from the menu.

4. The **Select extent** dialog will pop up. Use it to select the trees layer from the combobox, and click on **OK**. The coordinates of the layer extent will be added to the field.

5. Set 1000 as the horizontal and vertical spacing. These values have the same meaning as the search radius in the case of heat maps. Grid spacing determines how many cells will be in a grid and how smooth the resulting map will be. A smaller spacing produces smoother results, but very small values prevent us from identifying any distribution patterns. Note that the spacing should be specified in the same units as used by layer; for example, if a layer CRS uses feet as units, then the spacing should be in feet too.

6. Finally, in the **Output** field, specify the path where the resulting grid will be saved and click on **Run** to create the grid. When the algorithm execution completes, a new polygonal layer will be added to QGIS.

Counting points in grid cells

To calculate the number of features inside each grid cell, we can use fTools or Processing core plugins. The latter is more flexible and allows us to automate tasks, as described in *Chapter 8, Automating Analysis with Processing Models*. We will use the **Count points in polygon** algorithm from the Processing framework.

 There is also a **Count unique points** algorithm in Processing that lets us count only points with unique attributes in the selected field. This may be useful when a more precise analysis is needed, for example, for mapping diversity.

In the Processing toolbox, find the **Count points in polygon** algorithm by typing its name in filter field at the top of the toolbox. Double-click on the algorithm name to open its dialog, and you will see this:

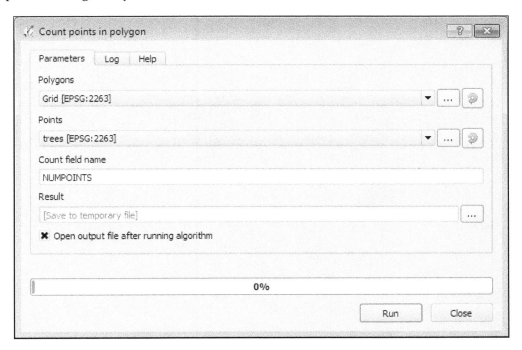

Select the grid layer created in the previous section as **Polygons** and the `trees` layer as **Points**. Enter **Count field name** or leave it unchanged. Don't forget the 10-character limitation for the shapefile field names! Finally, in the **Result** field, specify the location where the resulting layer will be stored, and click on the **Run** button to start the analysis. Keep in mind that this may take some time, as the `tree` layer contains many features and the grid size is relatively small.

When an algorithm execution completes, a new grid layer, containing the field with the number of points in each cell, will be added to QGIS. We can safely remove the original grid layer from QGIS and from the filesystem, as it is no longer needed.

For better visual representation of our data, we apply a graduated renderer to style cells according to the number of features in them. If necessary, go back to the *Developing styles for vector layers* section of *Chapter 2, Visualizing and Styling the Data.* The result may look like this:

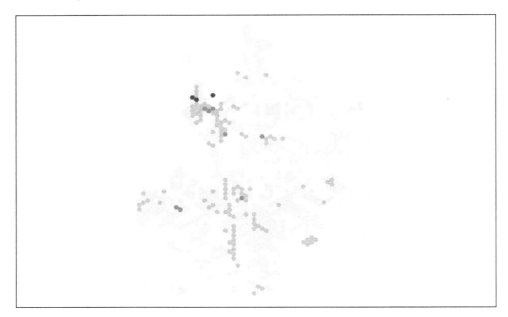

Removing redundant data

If we look carefully at our grid layer and its attribute table, we will see that some grid cells are empty; there are no points in them. Such cells displayed with a very light green color in the preceding screenshot. It is clear that the empty cells are much more than the nonempty cells. Of course, we can simply hide such empty cells by assigning to them the same color as the map background, or by removing them from the renderer (assigning an empty style). But, is it better to remove them completely and reduce the file size?

There are two possible ways to remove empty cells: manually or by using one of the existing tools from the Processing toolbox. Each method has own advantages.

First, you will learn how to remove redundant data manually. This method can be used when you need to process only one layer and don't want to create any temporary intermediate files. Also, it allows you to easily examine the features that will be removed. Let's do it:

1. Select the LAYER layer from the QGIS layer tree. Then, click on the **Select features using an expression** button in the **Attributes** toolbar to open the **Select by expression** dialog.

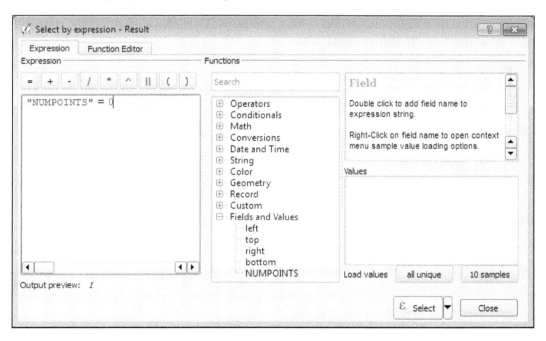

2. In the functions tree, under the **Fields and Values** group, find the NUMPOINTS field and double-click on it to add its name to the expression.

3. Then, click on **=** to add the equal to operator to the expression, and type 0 after it. The final expression will look like **"NUMPOINTS" = 0**.

4. Click on **Select** to select all the features that match your condition. The selected features will be highlighted in yellow so that you can check whether they are what you want.

5. Now, close the **Select by expression** dialog and open the layer attribute table by clicking on the **Open Attribute Table** button in the **Attributes** toolbar.

6. Toggle editing mode by clicking on the corresponding button in the attribute table dialog, or by simply pressing *Ctrl + E*.

7. To remove the selected features, just press the *Delete* key or click on the **Delete selected features** button.

8. Finally, turn editing mode off by pressing *Ctrl + E* again. Now our layer contains only nonempty cells and is much smaller in size.

Deleting redundant data with Processing requires fewer steps than removing redundant data manually, but instead of updating the existing layer, it creates another file. The Processing framework is useful when you need to process many layers at once or want to automate some operations. To remove empty cells, we can use the **Extract by attribute** algorithm.

 In our case, the comparison condition is very simple, so we use the **Extract by attribute** algorithm. When a more complex comparison condition that includes several attributes or some calculations is needed, it is better to use the **Select by expression** algorithm in combination with **Save selected features**.

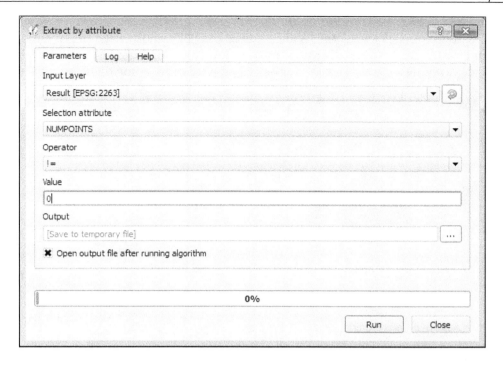

Here are the necessary steps:

1. Find the **Extract by attribute** algorithm by typing its name in the filter field, and double-click on its name to open its dialog.

2. In the **Input layer** combobox, select LAYER.

3. In the **Selection attribute** combobox, select the NUMPOINTS field, which stores the point count in each cell.

4. In the **Operator** combobox select the not equal operator !=.

5. Enter 0 in the **Value** field.

6. Specify the name of the output file in the **Output** field.

Now, we can press the **Run** button to start processing. When it is done, the new layer, with empty cells removed, will be added to QGIS, and we can remove the original layer. The new layer is much smaller, but still contains all of the information. Also, we can now set up styles more precisely and identify some distribution patterns that were not visible earlier.

If we overlay a layer that has a street network with this density map, we can easily identify that the greenest street is the Ocean Parkway.

The result looks great, but performing all of these steps manually is not very comfortable. Also, there are a number of intermediate layers that can, and should, be removed. Fortunately, QGIS allows us to automate such operations with the Processing Graphical Modeler. Refer to *Chapter 8, Automating Analysis with Processing Models*, to learn how to create models and use them.

Summary

In this chapter, we covered the analysis of dense-point datasets. You learned how to create raster heat maps with the help of the QGIS Heatmap plugin. Then, we covered analysis of heat maps and data extraction from them.

You also familiarized yourself with an alternative and popular technique for displaying dense datasets, called binning, and you learned how to perform hexbinning (a kind of binning technique) in QGIS using the Processing framework.

6

Answering Questions with Visibility Analysis

Visibility analysis is a valuable part of GIS analysis that answers questions such as "What can be seen from this location point?" In this chapter, you will be exposed to the basics of visibility analysis through defining the roof that provides the most scenic view of an area, and where the viewing platform can potentially be located. Throughout the chapter, you will learn how to do the following:

- Prepare data to represent urban landscape features
- Select potential observation points
- Compute the viewsheds to find the most scenic points
- Present the results in a 3D scene

Before we start, we will explore some essential principles of viewshed analysis.

The basics of visibility analysis

The aim of visibility analysis is to produce a coverage of an area that can be seen from a specified location. This coverage is called a **viewshed**, which is why the terms **visibility analysis** and **viewshed analysis** are used interchangeably. To perform a simple visibility analysis, you need to define at least two components:

- **Observation point**: A point that represents an observer's position and for which visibility is being analyzed
- **DEM**: This represents irregularities in the earth's surface and is used to examine visibility along the line of sight

The idea underlying the process of this analysis is to compare the height of the observation point against the height of earth's surface point along the given line of sight. If the height of the surface point is less than that of the observation point, then it will be seen from the current position; if it is higher, the visibility line will be blocked. Similarly, all points within a certain radius are compared against the observation point and divided into the two categories, as follows:

- Visible, with the height lower than the observation height and located below the visibility line

- Invisible, with the height higher than the observation height and located above the visibility line

The main output of visibility analysis is a binary true/false viewshed coverage that usually includes visible points and excludes invisible points. Depending on the algorithm used and the capacity of GIS, this basic analytical approach can be modified and improved by the following features and outputs:

- Instead of a single observation point, multiple points and even lines can be used

- Visibility relationships between several objects can be analyzed, and then intervisibility coverage is produced as the output

- Conversely to the assessment of a viewshed for the current observation point, the earth's surface can be analyzed to provide information about points and areas from which a certain object can be seen

- A line of the horizon can be modeled as the cumulative edge of visibility zones

- Views can be bound by a horizontal or vertical viewing angle or azimuthal values that limit extension of visibility lines

- The earth's curvature and atmospheric refraction can be taken into account to simulate more realistic results

The most typical practical application of visibility analysis is in placement of communication towers, where instead of visibility, signal penetration is modeled. Viewsheds are also of great use in territorial and urban planning. For example, some features such as factories or landfills are expected to be hidden from the human' eye because of their unsightly appearance. In this case, the area undergoes a visibility analysis to find places from where such kinds of objects can't be seen. The more the blind spots associated with a certain point, the better it is for the object's location. Furthermore, instead of hiding some objects, viewsheds are used to detect scenery points that provide the most spectacular view of an area. This knowledge is then used for optimal placement of the sight places, which is what we are going to do in this chapter.

Step 1 – converting a buildings' vector layer to raster

We are going to deal with a highly urbanized landscape, whose primary features have been greatly transformed by humans. In the context of visibility analysis, this means that lines of sight are blocked mainly by buildings, while the original relief features play only a minor role. In our dataset, we have two layers that describe the area of interest, exterior, and appearance:

- `lidar_dem`: This represents the bare earth only, and provides a description of the general relief

- `building_footprints`: These are polygons that depict all buildings, and they also contain information about a building's height in feet above the bare earth's surface in the `height_roo` attribute field

We need to add their values in order to obtain a meaningful result, but these layers use different data models: DEM is a raster coverage, and building footprints is a vector polygon layer. That is why before adding them, we should first take layers to a single common data model. In GIS, calculations for coverages are typically performed using raster algebra, which means that the `building_footprints` vector layer should be converted into a raster, or rasterized. Additionally, if we want this raster layer to contain data about building's height, the `height_roo` attribute field should be used as the provider of Z values.

It is also important that a newly created building footprint's raster has the same extent and resolution as `lidar_dem`. To get this information about `lidar_dem`, go to **Layer Properties | Metadata**. At the bottom of the window, you will see the **Properties** scrolling window section, which contains all of the information important to know in order to work with rasters, namely **Band 1** (statistics), **Dimensions** (number of rows and columns), **Origin** (coordinates of the bottom-left corner), and so on. The parameters of particular interest to us are as follows:

- **Pixel Size**: This contains the vertical and horizontal lengths of the pixel's sides. Usually, the values are equal, which means that the pixel is a square; but sometimes, it can be rectangular, with unequal values.

- **Layer extent (layer original source projection)**: This represents the minimum and maximum horizontal (x) and vertical (y) values of the raster extent, that is, the raster bounding box's northern, southern, eastern, and western limits in the layer's original coordinate reference system.

The following screenshot shows the described parameters for the `lidar_dem` layer:

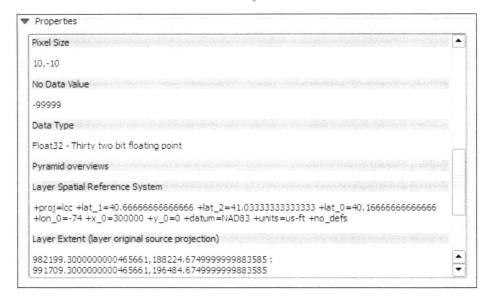

To rasterize a layer, go to **Raster | Conversion | Rasterize (Vector to Raster)**. In the dialog window, adjust the following parameters:

1. **Input file (shapefile)**: This is a shapefile to be rasterized. Select `building_footprints` from the drop-down list.

2. **Attribute field**: This defines an attribute field with the values to be burned into an output raster. Select `HEIGHT_ROO` from the drop-down list.

3. **Output file for rasterized vectors (raster)**: There are two possible options in this field. You can select an already existing file, and patches of vector geometry footprints with the selected values will be burned into it. It is very convenient if you have, for example, some gaps in the original raster layer and want to fill them with some values from the vector layer (such as elevation and surface level of water bodies). But we have to deal with the buildings' relative height above the bare earth level. This means that, in order to summarize them with DEM, building footprints should be rasterized as an individual layer. Navigate to your working directory and type `building.tif` as a new layer name. You will be shown this message: **The output file doesn't exist. You must set up the output size or resolution to create it.**. Click on **OK** and proceed to the following stage:

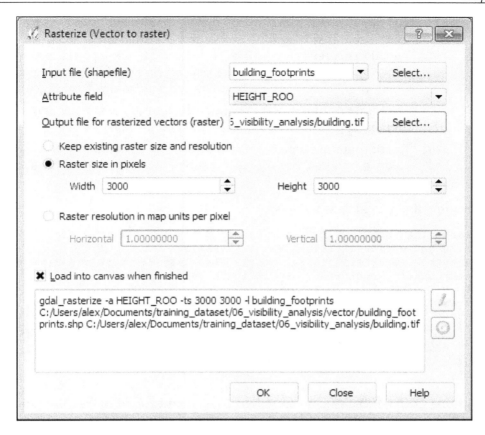

4. In the previous steps, you defined some major options. As we need to create a raster with exactly the same extent and resolution as `lidar_dem.tif`, we will use advanced options by editing the `gdal_rasterize` command-line parameters:

    1. Click on the **Edit** button to make the line editable. All the previous options will be deactivated and grayed out.

    2. Delete the `-ts 3000 3000` option, as we are going to specify a resolution parameter that makes raster width and height parameters that are defined by `-ts` meaningless.

    3. The `-init 0` parameter defines the initial values of the output raster. It creates an empty raster with initially predefined values, and then vector values are burned into it. In our case, this means that building-free areas will be assigned zero values.

4. The target extent is defined by the `-te` parameter described by space-separated bounding box values (**Xmin**, **Ymin**, **Xmax**, and **Ymax**) in georeferenced units. If you want to use a predefined extent, just copy its values from a piece of correspondent raster metadata. Don't forget to replace the commas and semicolons with spaces.

5. The target resolution (`-tr`) parameter defines the vertical and horizontal pixel size in units specified by the layer's CRS.

6. The output data type is defined by the `-ot` parameter, whose default value is `Float64`, but we replace it by `Float32`, which is the same as that for `lidar_dem.tif`.

The resulting line will look as follows:

```
gdal_rasterize -a HEIGHT_ROO -l building_footprints -init 0 -te
982199.3000000000465661 188224.6749999999883585
991709.3000000000465661 196484.6749999999883585 -tr 10 10 -ot Float32
fullpath/building_footprints.shp fullpath/building.tif
```

> To convert a vector to a raster, we have used a tool called `gdal_rasterize` from the **Geospatial Data Abstraction Library** (**GDAL**). You can read an extended synopsis of the `gdal_raserize` parameters and their values from `http://www.gdal.org/gdal_rasterize.html`.

After hitting **OK**, the resulting raster will be created and added to the map canvas. If you click on any pixel with the **Identify Features** tool selected, a height value of the building will be displayed, and if you click on an empty space, it will show `0`. The layer itself will look similar to the following screenshot:

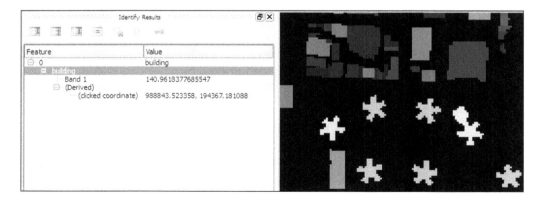

Step 2 – combining the DEM and buildings layers

Now, we need to combine the two layers into a single layer by adding their values. **Raster algebra** is a common approach to overlaying a raster layer (or layers), combining their values using algebraic operations (addition, subtraction, multiplication, division, and so on), and calculating values for a new raster layer. Modern GIS software uses so-called raster calculators that help create, validate, and apply raster algebra expressions to raster datasets.

QGIS also has its own raster calculator located at **Raster | Raster Calculator**. The **Calculator** dialog window consists of the following parts:

- **Raster bands**: In the top-left corner of the window, you can see the list of all rasters available in the project. The raster band number is separated from its name by an @ sign. If the layer is a single-band raster, then only name@1 appears in the list, but if it is a multiband raster, then name@1, name@2, and so on up to name@n (where *n* is the total number of bands in the multiband raster) will be shown for each band separately.

- **Result layer**: In the top-right part of the calculator window, you can adjust the output raster properties. The **Current layer extent** button sets up an output extent from the raster layer that is currently selected in the **Raster bands** list on the left side. It is very useful when you are working with layers that have different extents, and allows you to decide which extent to use.

- **Operators**: In the central part of the window, there are various operator buttons that can be used to construct expressions. Click on the relevant button to add an operator, or type it manually.

- **Raster calculator expression**: In the bottom part of the calculator, there is the expression window, where the calculation formula to be used is displayed. Double-click on the relevant layer to add it to the expression (it will be shown in double quotes). Note that for the expression, we type only the right side of the calculation formula, which comes after the equal to sign. Under the expression, you will see the **Expression valid/ invalid** message change dynamically, and this helps you control the correctness of the constructed formula.

In the following screenshot, you can see that we apply a simple addition formula (`"building@1" + "liadar_dem@1"`) to combine rasterized buildings' heights and DEM layers:

After you've clicked on the **OK** button, the calculation is performed and a newly created `urban_surface` raster appears in the **Layers** panel. Now, we can check its values with the **Identify features** tool:

1. Uncheck all unnecessary layers except those from `urban_surface`, `building`, and `lidar_dem`. Maintaining their order simplifies the interpretation of **Identify Results**.

2. Go to the **Identify Results** tab. From the **Mode** drop-down list, select **Top down**. The identified values will be shown for all the active layers in order, from top to bottom.

3. From the **View** drop-down list, select **Table**. The identified values will be organized into a simple table, as shown in this screenshot:

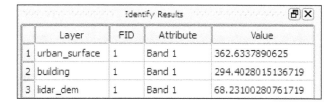

Now, by clicking on any point in the map canvas, you can explore values and make sure that they have been added properly.

Step 3 – defining observation points

Observation points play an important role in visual exploration of modern cityscapes. Usually, they are represented by the highest points in an area that provide the most spectacular views. Therefore, viewing platforms in the city are usually located on rooftops. In this section, we are going to create a layer that contains several prospective viewing platforms through the following steps:

1. Creating an empty vector layer
2. Populating it with some points that represent the highest buildings
3. Providing them with the information on height from the `buildings_footprint` layer

Creating an empty vector layer

To create a vector layer, go to **Layer | Create layer**. There are three options available here:

- **New Shapefile Layer**: Also accessible by the *Ctrl + Shift + N* keyboard shortcut, this creates a new empty shapefile layer.

- **New SpatiaLite Layer**: Also accessible by the *Ctrl + Shift + A* keyboard shortcut, this creates an empty SpatiaLite layer within a specified SpatiaLite database (by default, a currently connected database is specified).

- **New Temporary Scratch Layer**: This creates a temporary layer that can be used and analyzed like any other layer within a working session, but the layer will disappear if it is not saved before closing the project. These layers are meant to be drafts, and using them for testing purposes prevents cluttering your project. This is the option we will use to create an empty layer.

Similarly, you can use a relevant button from the **Manage Layers** toolbar and the small triangle beside it to get access to the options, as shown in the following screenshot:

After selecting **New Temporary Scratch Layer**, you will see this dialog window:

Consider the options in the **New Temporary Scratch Layer** window, as follows:

- **Layer name**: Type in a name or leave the default as it is, because we are going to use the layer for temporary work and its name really doesn't matter.
- **Type**: This defines the geometry type. As we are going to set up observation points, the default **Point** option is suitable.
- **Selected CRS**: A default projection is selected, but you need to set up a projection of your project in order for the layer to be processed correctly. To do so, select the **Project CRS** option from the drop-down combobox.

After clicking on **OK**, a **New scratch layer** will be added to the **Layers** panel. By default, editing mode for the layer is activated (you can see it as a little pencil drawn above its marker symbol beside the layer's name) as shown in the following screenshot. Now we can proceed to the next stage and add some points to it.

Populating a layer with points

Activate the **Digitizing** toolbar, if it has not already been done. You can do this by going to **View | Toolbars | Digitizing**, or simply by right-clicking somewhere on the toolbar's panel and activating the relevant toggle. The toolbar contains the primary digitizing tools, and looks like this:

This panel's buttons provide access to the following options (from left to right):

- **Current edits**: This maintains edits within a current editing session for the selected layer (or layers). Click on the little black triangle in the bottom-right corner of the button to access the **Save**, **Rollback**, or **Cancel** options for your edits. Note that these options are available until the **Save Layer Edits** button isn't clicked on.

- **Toggle Editing**: This button activates/deactivates the editing mode.

- **Save Layer Edits**: This saves all the current edits without exiting the editing session. After clicking on this button, edits cannot be rolled back. It is inactive when there are no edits to save.

- **Add Feature**: Use this button to draw new features on the layer. Its appearance depends on layer's geometry type. Currently, it shows points because you are editing a point layer.

- **Move Feature(s)**: When this button is selected, you can move one or several features. It is important to note that if you are going to move multiple features, they should have been previously selected.

- **Node Tool**: This tool is used to add, remove, or move vertices of geometric features such as lines and polygons. It is unnecessary for a point layer and is therefore inaccessible.

- **Delete Selected**: This deletes the previously selected feature (or features). Until the edits are saved, they can be rolled back through the **Current Edits** button options.

- **Cut/ Copy/ Paste Features**: Similarly to other kinds of software, you can use these buttons to cut or copy features onto the clipboard and then paste them. Features must be previously selected if you want to apply these options. While **Cut** and **Paste** are available only in editing mode, **Copy** can be used with any other layer, and is of great use when moving features between layers.

You can also use more complex digitizing tools and options, such as feature rotation, simplification, adding parts and rings, and so on. These are available from the **Advanced Digitizing** panel, as shown in the following screenshot:

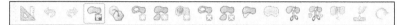

 Moreover, the panel provides access to input tools that are similar to those used in **computer-aided design** (CAD) systems. These tools allow digitizing with precise numerical values of coordinates, distances, and angles; control segments; parallelism; and perpendicularity, as shown in the following screenshot:

Now that you are familiar with the **Digitizing** toolbar, **Toggle editing**, and using **Add Feature**, click on different places to locate some observation points. Remember that these points should be placed on the footprints of the highest buildings. It is recommended to use the previously created `urban_surface` raster as a background, as it helps locate the points properly. Now, add several points (five to seven are enough) by clicking on the map canvas. Click on the **Toggle editing** button to exit editing mode. Click on the **Save** button when you are asked about edits. As a result, your map will look similar to this screenshot:

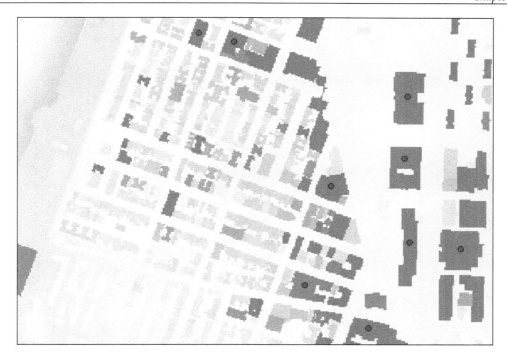

Providing points with height values

As we are going to locate viewing platforms on the rooftops, taking into account
the height of the building is of great importance to obtain realistic modeling results.
Instead of looking for the necessary building and its height and adding the height
value to the visibility points manually, we will automatically join attributes, taking
into account the locations of the points. This operation—when you merge attributes
from two different layers based on their spatial relationship—is very common in GIS
and is called a **spatial join**.

Go to **Vector | Data Management Tools | Join Attributes by Location**. In the dialog
window, the following options are present:

- **Target vector layer**: The layer to which attributes will be joined; in our case,
  it's **New scratch layer**.

- **Join vector layer**: The layer from which attributes will be taken; in our case, it is `building_footprints`.

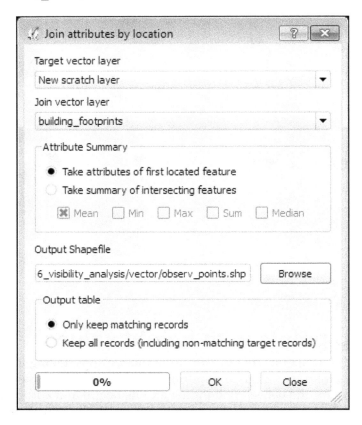

- **Attribute Summary** deals with multiple attribute values available for joining. The **Take attributes of first located feature** option joins a single value, and this is the default action that we select. Optionally, you can activate **Take summary of intersection features** and select one or several summarizing functions (**Mean**, **Min**, **Max**, and so on).

- **Output Shapefile** provides the path to the resulting vector layer. Click on the **Browse** button to select the necessary directory and type a name, for example, `observ_points`.

- **Output table** is responsible for the number of records in the resulting attribute table. We leave the **Only keep matching records** default option as selected because we only need records of buildings that match with the observation points.

After clicking on **OK** and completing the spatial join, you will be informed that a new layer has been created and asked, "Would you like to add the new layer to the TOC?" Click on **Yes**, and the layer will be loaded and appear in the map canvas. Click on the **Close** button to exit the **Join attributes by location** dialog window.

Now, if you open the `observ_point` attribute table, you will see that all the attributes from matching buildings in the `building_footprint` layer have been joined to the points. This tool may be of great use when working with a large number of entities.

Step 4 – creating viewshed coverages

In this section, we will apply the advanced visibility analysis tools that are available from the **Viewshed Analysis** plugin. Install it as described in the *Extending functionality through plugins* section of *Chapter 1, Handling Your Data*.

After the plugin has been installed, you can get access to its functionality by going to **Plugin | Viewshed Analysis**. The **Advanced viewshed analysis** dialog window consists of three tabs. The **General** tab provides access to all the available analysis and output options. The **Reference** tab contains brief descriptions of the options and a link to the project's homepage, where you can read detailed information about the algorithm implemented in the plugin and report bugs, if any. The **About** tab contains brief information about the author and the plugin's homepage link.

Use the following options under the **General** tab to create viewshed coverages, as shown in the following screenshot:

- **Elevation raster**: Select the `urban_surface` raster to represent the earth's surface and buildings on it.

- **Observation points**: Select the `observ_points` point layer.

- **Output file**: Click on the **Browse** button, navigate to the working directory, and type a name for the output raster layer. Note that one output raster file will be created for each observation point, and the typed filename will be used as a template that will be accompanied by the output coverage type (viewshed, intervisibility, invisibility, and horizon) and index number. For this reason, we call the output layer `coverage`, presuming that the explanatory parameters (coverage type and point number) will be added automatically.

  Our goal is to find the point (or points) of the most scenic view, so there is no need to analyze the points' intervisibility. Hence, the **Target points (intervisibility)** field is omitted.

- **Search radius**: This is the size of the area observed around a point in the layer's measurement units (feet in our case). The higher the value, the longer the time taken to produce coverages. Accept the default value of 5000.

- **Observer height**: This defines the height of an observation point above the earth's surface. Usually, the higher the point, the better the observation. Instead of the predefined height, select HEIGHT_ROO from the field drop-down list.

- **Target height**: This is necessary when you want to know whether any target objects of a specific height are visible from the current position. We omit this option because we are more interested in the general visibility than in the visibility of specific objects.

- **Adapt to highest point at distance of**: This option finds the highest point in the observer or target vicinity within a certain area, defined in pixels. It may be of great use when your observation points provide approximate locations, and you need their height values to be adjusted to the highest point within the surrounding area. As we already have the exact height values, this option will be omitted.

- **Output**: This can be represented by the following coverages:
  - **Binary viewshed**: This is a simple raster coverage where all pixel values are assigned 0 (invisible) or 1 (visible). Activate this option to create viewsheds.

  - **Invisibility depth**: This measures the size an object should attain in order to become visible if placed in an area which is out of view. In other words, it defines the number of height units required for an object to become visible within a given radius. The output raster is a kind of inverted visibility raster where all visible pixels are assigned 0 as the value and all invisible pixels are assigned negative height values (the lower the value, the more invisible an object).

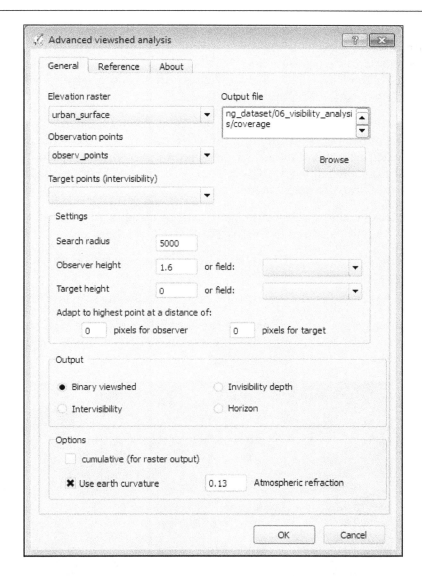

- ° **Intervisibility**: The output will be a shapefile that represents a network of visual relations between the observation points. In the output network, points that can be visible from each other are connected.

- ° **Horizon**: The output raster coverages represent the horizon line or the edge of the visibility area.

Optionally, instead of multiple coverages, a single **cumulative** raster that summarizes raster coverages to generalize analysis results can be created. Then, 0 values are assigned to invisible pixels, and positive values are assigned to visible pixels (higher values correspond to pixels visible from multiple observation points).

- **Use earth curvature option**: This takes into account the earth's curvature and the effects of refraction of light when traveling through the atmosphere. Activate this option and accept the default 0.13 refraction value. After all the options are adjusted, click on the **OK** button and wait for the results to be loaded into the map canvas.

Similarly, you can create the **Horizon** individual coverages. As a result, a total of 18 raster coverages will be added to the map canvas:

- coverage_N_Binary: A binary visibility raster where *N* is the number of observation point
- coverage_N_Horizon: A raster that represents the horizon line for the point *N*

In the following screenshot, you can see the example combination of binary visibility (on the left) and the horizon line (on the right) coverages for the same observation point:

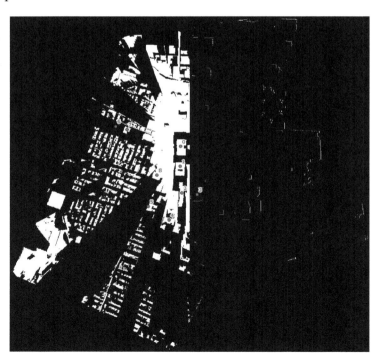

Step 5 – finding scenic points

Now, we should go through all the observation points and corresponding coverages, adjust their visualization options, and select the point (or points) that provides the best view of the area of interest.

First of all, we need to enumerate the points properly in order to be able to distinguish between them. Follow these steps to do so:

1. Select **Open Attribute Table** from the `observ_points` by right-clicking on the contextual shortcut, or click on the correspondent button ▦ from the **Attributes** toolbar.

2. Select **Open Field calculator** ▦ from the table toolbar or use the *Ctrl + I* keyboard shortcut.

3. In the **Field calculator** dialog window, activate the **Update existing field** option, and make sure that the `OBJECTID` attribute field is selected from the drop-down list, as shown in the following screenshot. This field already contains values that were joined from the `building_footprints` layer. These values are independent of the point number, and this is why we are using the update option to change the existing values to consistent values.

4. In the **Expression** tab, expand the **Record** group, which contains functions that operate on record identifiers. Select and double-click on the **$id** function to be added to the **Expression** window. This function returns the feature ID of the current row, which will help us identify the points and the correspondent coverages properly.

Click on the **OK** button. **Edit mode** will be automatically turned on and the records' values will be updated.

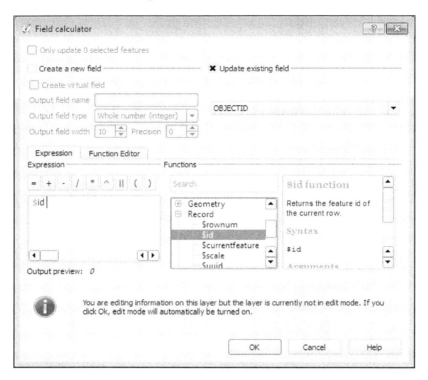

5. **Toggle editing mode** from the **Table** toolbar or use the *Ctrl + E* keyboard shortcut to exit editing mode. You will be asked whether you want to save the changes made to the observ_points layer. Click on the **Save** button and close the attribute table.

6. Double-click on the observ_points layer to open its **Layer Properties** dialog window. In the **Labels** section, activate the **Label this layer with** option and select the OBJECTID field for labeling. Adjust any other labeling options, if needed, and click on **OK** to exit the window. When the observation points are labeled, it will be much easier to analyze them and their viewsheds.

Now, we will adjust the symbology of the visibility layers to simplify their interpretation. Let's start from the very beginning — point number 0. Follow these steps to achieve meaningful results:

1. Activate the `coverage_0_Horizon` and `coverage_0_Binary` layers, and deactivate all other visibility layers. Make sure that the layers lie under `observ_points`. Activate any other layers that might be of interest when analyzing visibility (for example, `urban_surface`, `roads`, and so on), and make sure that they underlie visibility rasters.

2. Adjust the `coverage_0_Horizon` layer's style. This layer contains only two values: `1` represents the horizon line (that is, the edge of the visibility area), and `0` represents all other areas. Double-click on its name in the **Layers** panel to open the **Layer Properties** dialog, go to the **Style** section, and adjust the following parameters:

   ° From the **Render type** drop-down list, select **Singleband pseudocolor**.

   ° Set **Color interpolation** to **Exact**, as we have only two values and want them to be assigned to the selected colors.

   ° Use the ⊕ button to add the necessary values manually. Enter `0` and double-click on its **Color** sample. In the **Change color** window, select black from **Standard colors**, and using the **Opacity slider**, set the transparency to `25%`. Click on the **OK** button to come back to the main styling options.

   ° Again, add one more row for the value of `1`, and assign it a bright, contrasting color to make the horizon line distinguishable on the map. Type in the explanatory names for values in the **Label** field. Your color table should look similar to the following screenshot:

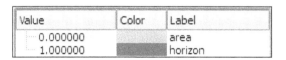

Value	Color	Label
0.000000		area
1.000000		horizon

3. Open the `coverage_0_Binary` properties dialog by double-clicking on the layer name in the **Layers** panel, and go to the **Transparency** section. We will use **Custom transparency** options and **Transparent pixel list** to adjust the pixels' transparency properties according to their values. The binary visibility raster contains only two values: `1` for visible areas and `0` for invisible areas. To add these values to the list, click on ⊕, the **Add values manually** button. A new row will appear in the list. There, you should manually enter the **From** and **To** values, and adjust or accept the **Percent Transparent** value, which is set to `100` by default. As we are going to use values instead of ranges to set the transparency, the **From** and **To** entries will be identical:

   ° In the first row, enter `1` and accept the 100% transparency.

     °   Click on ⊕ again to add the next row, enter 0 in **From** and **To**, and set the transparency to 25%. As a result, your **Transparent pixel list** will look like this:

	From	To	Percent Transparent
1	1	1	100
2	0	0	25

After making all the necessary adjustments, click on the **OK** button to apply them.

4. Apply the same styling options for the layers of the other points. Instead of working on every raster individually, right-click and navigate to **Styles | Copy Style** to replicate the preliminary configured visualization options from `coverage_0_Horizon` and `coverage_0_Horizon`. Then, go to **Styles | Paste Style** to insert them into the appropriate layers.

If the coverages are visualized properly, we only need to visually interpret them and make a decision about the best points after considering several criteria. First of all, a point should provide a view for the vast areas, and this can be analyzed through binary coverages; the more the area open to the observer, the better. Secondly, the horizon line should be clear, solid, and not disrupted by artifacts, if possible. The integrity of the visibility edge ensures panoramic views, and it can be analyzed through horizon coverages; the straighter and simpler the line, the better.

Finally, you should always take into account the local features represented by sights, open spaces, and any other possible points of interest. Regarding New York city's Brooklyn borough, there are a few remarkable features that you have probably heard about: the Brooklyn bridge and the waterfront area where the Brooklyn Bridge Park is located. This area is famous for its beautiful sunsets and awe-inspiring views of Manhattan and the East River. Taking into consideration all of these criteria, point number 7 would probably be the best choice. Not only does it provide views of the southern and northeastern parts of the area, but it also encompasses the waterfront area almost entirely, as shown here:

When the winner among the best view is chosen, we can proceed to the following step and visualize the results of the analysis. Instead of representing them in a conventional cartographic way, we will use the power of three-dimensional visualization, which is of great use when dealing with urban landscapes.

Step 6 – styling the results in 3D

In this section, we will represent our data in a 3D scene using the impressive capabilities of the **Qgis2threejs** plugin. Install it as described in the *Extending functionality through the plugins* section of *Chapter 1, Handling Your Data*. This plugin relies on the three.js library. It allows us to export terrain data, the map canvas image, and vector data straight to a web browser. As a result, you can view and explore exported objects as a 3D scene on web browser that supports WebGL.

After the installation is completed, the plugin is available from the **Qgis2threejs** menu under **Web**. There are two submenus available:

- **Settings**: This is responsible for defining a browser that will be used to open generated 3D scenes. These settings should not be changed if you want your default browser to be used.

- **Qgis2threejs**: This is the main window of the plugin, and it provides access to its general functionality. In the left part of the window, you can see the list of all available control parameters and the project's layers. The right part of the window provides access to their settings and changes interactively, depending on the item selected on the left.

First of all, you need to adjust the map layers that will be used to generate an image draped over the terrain. For this example, we activate the following layers ordered from bottom to top: `water_area`, `parks`, `roads`, and `coverage_7_Binary`. You can always add more layers or even use the predefined WMS/ WFS or OpenLayers plugin coverages, for example, OpenStreetMap. These active layers will be used to provide background terrain.

 In the following sections, we are going to explore only the basic settings of the plugin that are necessary for generating a meaningful result from the training data. Extended help and descriptions of the parameters are available at `https://github.com/minorua/Qgis2threejs/wiki/ExportSettings`.

Working on the general settings of a 3D scene

Once all the necessary layers are active, go through the following options to adjust the general settings of the scene:

- **Template file**: Here, you can select different output templates from the drop-down list. Use the default **3DViewer(dat-gui).html** option, as it not only generates a 3D scene but also adds a control parameters panel to regulate the visibility and opacity of the layers and vertical movement of a custom horizontal plane.

- **World**: This item is responsible for the general appearance of the scene in the 3D world. The parameter we are especially interested in is **Vertical exaggeration**, as it defines the complexity of relief appearance. The bigger the value, the more emphasized terrain will reflect the relief features. Also, this value affects the Z positions of all vector 3D objects and 3D object heights of some object types with volume (points and extruded polygons). The default value is set to 1.5, but our area of interest has a plane relief with smoothly changing values, so we enter 2 to make changes in the height more obvious.

- **Controls**: There are two available control choices, accompanied by descriptions. Leave the default `OrbitControl.js` unchanged.

- **DEM**: From the **DEM** layer drop-down list of available rasters, select `lidar_dem`, which will be used to provide actual information about heights of the area and generate a terrain of predefined vertical exaggeration. There are several sections that provide parameters to regulate DEM appearance:

    ◦ The **Resampling** block is responsible for the generated surface resolution and provides various options for its adjustment. For the purpose of this tutorial, we will simply move the slider to the fourth tick, which gives an output resolution close to 400 x 400 pixels and resamples the original DEM to approximately 26.8 feet. This approach is fast and produces a lightweight output, but resampling values affects the DEM's resolution, not the texture draped over it. If you want to produce more sophisticated and high-resolution results, read about the advanced settings, but bear in mind that high-resolution output takes more time to export, and also that the scene itself will require more computer resources to be responsive and rendered fast.

    ◦ **Display type**: Accept all the default settings that presume use of **Map canvas image** as a texture.

    ◦ **Sides and frame**: Accept the default **Build sides** option. It adds sides and a bottom to the DEM.

- **Additional DEM**: this option is might be of great use if you want to combine two rasters that contain some Z values (for example, absolute height and slope, heat maps and so on). As we are mainly interested in buildings and binary visibility rasters, this option is omitted.

Adjusting 3D visualization of the observation points

Expand the **Point** item on the left side of the window to see all the available point layers, and activate `observ_point`. In the right part of the window, adjust the following settings:

1. For **Object type**, accept the default **Sphere** value. The observation points will be represented by spheres in the 3D space.

2. Set **Z coordinate Mode** to **+"HEIGHT_ROO"**, which means that the height of a point will be obtained from the following formula: *z = elevation at vertex + field value + addend = DEM + HEIGHT_ROO*. In other words, we only summarize the elevation of the relief and the height of the building at a certain point, without any other values (*addend* is set to 0 and excluded from the second part of the formula). As a result, the point will take into account the local relief and represent the real vertical position of the observer.

3. The **Style** section is responsible for the visual properties of the symbol:

   ° Set **Color** to **Random** from the drop-down list, and the spheres will be assigned a color randomly.

   ° Accept the default **Feature style** option under **Transparency**. It will make the points solid, not transparent.

   ° As we want all spheres to have the same size, we accept the **Fixed value** under **Radius** as it is. Type 50 in the **Value** field.

4. Now, let's consider **Features that intersect with map canvas extent**. With this default option, only the features that are displayed on the map canvas will be exported.

5. The **Attribute and label** section provides access to the labeling options:

   ° Activate **Export** attributes to enable labeling. All the attributes will be exported and shown when you click on the relevant object in the 3D scene.

   ° From the **Label** field drop-down list, select the `OBJECTID` field that will be used for labeling.

   ° The **Label height** option defines how high from the sphere its label will be shown. Here, we select the **Height from point** option from the drop-down list and set **Value** to 100.

The following screenshot shows how the window will look after all the parameters are defined:

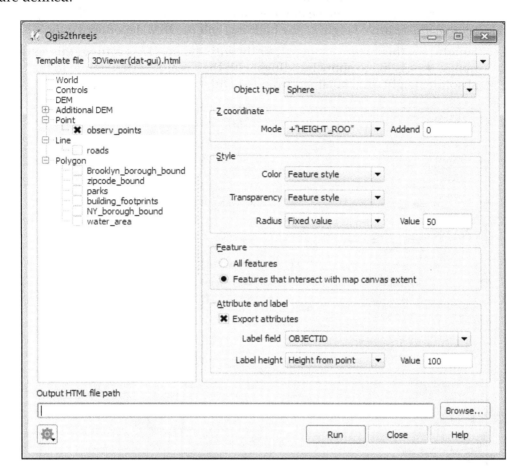

Adjusting 3D visualization of building footprints

Expand the **Polygon** item on the left side of the window to see all the available polygon layers, and activate `building_footprints`. On the right side of the window, change the settings to what is shown in the following screenshot:

- **Object type**: Set this to **Extruded**. This means that 2D polygon footprints will be represented as 3D parallelepipeds.

- **Mode**: Set this to **Relative to DEM**. This means that the height of the building will be obtained from the following formula $z = elevation\ at\ vertex + addend = DEM$. In other words, as we don't use any additional values and the height is set to 0, the relative height of the building will be DEM dependent only.

- **Height**: Set this to HEIGHT_ROO. Jointly with the previous setting, this setting makes it possible to show buildings with their real height, relative to the absolute height a.s.l. obtained from the DEM.

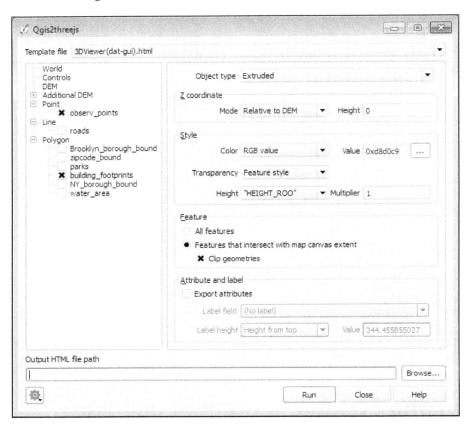

After all of these settings are done, click on the **Browse** button to specify the path and name for the output scene. Then click on **Run**. When the scene is generated, it will be automatically loaded and opened in your default web browser, where you can visually explore the results of the analysis using the previously defined controls, as shown here:

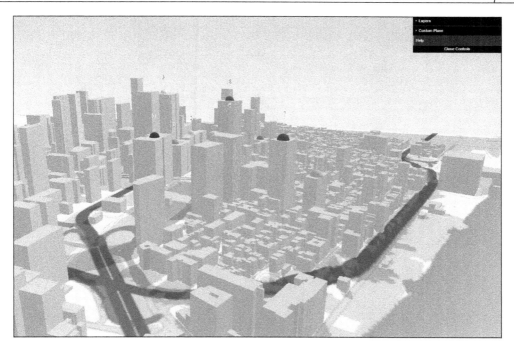

Summary

In this chapter, you were exposed to the basics of visibility analysis, including input data preparation, visibility point creation, and production and interpretation of visibility coverages in order to find the best observation point. Moreover, you learned how to represent your results through realistic 3D models, which are of great use in the visual exploration of urban data and the results of its analysis.

In the next chapter, you will learn how to define a perfect location by means of spatial analysis.

7
Answering Questions with Suitability Analysis

We live in a world full of various relationships that can be analyzed in functional, temporal, or spatial contexts. Spatial relationships are of particular interest when it comes to GIS, because here objects in space are represented in a way that facilitates explanations and analyses of their geographic relations. Suitability analysis is a fundamental part of GIS analysis that answers the question, "Where is the best place for something to be located?" In this chapter, you will be exposed to the fundamentals of suitability analysis through the search for the best place to live in. You will learn how to:

- interpret spatial relationships between objects
- express these relationships through spatial data
- analyze spatial data according to a set of predefined criteria
- overlap layers and interpret the results

Before starting with this chapter, some essentials of suitability analysis and types of spatial relationships between objects will be explained.

Basics of suitability analysis

Suitability analysis is recognized as a multi-criteria decision support approach. In other words, its main aim is to divide the area of interest into two categories based on a set of predefined criteria: appropriate for some kind of use (living, building, conservation, and so on) and inappropriate. A general approach that is used for suitability assessment is multiple layer overlays that support multi-criteria decisions. Depending on the data that represents suitability criteria and overlaid, there are two basic approaches available:

1. Suitability analysis with vector data. This primarily utilizes operations such as buffering and their sequential combination using vector overlay operations, such as clipping, intersection, and union.

2. Suitability analysis with raster data. This heavily relies on raster algebra, which is used to reclassify initial raster coverages and then combine them to produce binary or ranked suitability rasters. This approach is more flexible, as it makes it possible to produce multiple suitability classes and change the raster weight according to the importance of the factor it represents. As a result, the user is able to produce a compound result, but the workflow requires more effort connected to data preprocessing and consolidation decisions.

The following table summarizes the main geoprocessing stages, advantages, and pitfalls of suitability assessment with vector and raster data:

	Vector data	Raster data
Main geoprocessing and analysis operations	• Buffering • Clip overlay • Intersection overlay • Union overlay	• Vector data rasterization • Proximity raster creation • Raster reclassification • Raster algebra addition • Raster algebra multiplication • Raster algebra subtraction

	Vector data	Raster data
Advantages	• Workflow quickness • Workflow simplicity • Good representation of man-made features	• Simple data reclassification • Good representation of continuous features • Crisp and fuzzy classes are possible • Possibility to weigh different coverages according to their importance • Various assessments are possible, such as binary, ranked, and weighted
Limitations	• Provide only crisp classes • Usually provides binary (yes/no) assessments only	• Reclassification and ranking subjectivity • Workflow complexity

No matter what approach you will follow, the general suitability analysis workflow involves several common steps. We will now take a closer look at them to ensure better understanding of the suitability analysis system:

1. **Define the goal and objectives of your analysis**: The question to be studied is formulated in general, and its applied significance is determined, which will later become the set of suitability criteria. Some popular suitability applications include the following:

    ○ **Agriculture**: The appropriateness of the area for cultivation of certain crops is assessed.

    ○ **Retail**: The area is assessed from a marketing prospective—whether it will attract new customers or buyers, or not. This type of analysis is in great demand when selecting preferable shopping locations.

    ○ **Renewable energy**: Assessing land suitability for locations of wind power or solar power stations is a remarkable trend in the field of geospatial planning for sustainability.

    ○ **Nature conservation**: Conservancy needs are prioritized using habitat suitability modeling, and the area is divided into locations that are more or less valuable for certain species' survival and reproduction.

Generally speaking, the primary application of suitability analysis is in the field of land use planning, aimed at reasonable prioritization of various types of human activities within limited space and natural resources.

2. **Analyze the available data and define its relevance to the goals and objectives**: Data's relevance to the goal and objectives is defined. Current data availability and future data requirements are analyzed, especially the knowledge of whether the current data derivatives can be used for analysis or not. For example, if we have to analyze suitability for agricultural needs, a DEM can be a great source. It provides not only basic information about the relief, but also some useful derivatives, such as slope and aspect. The main outcome of this stage is a list of primary data sources and their potential derivatives.

3. **Define the criteria of analysis**: This is the most important stage, where the objectives of analysis are described as clear numerical criteria based on relevant data. Descriptive objectives are translated into the language of GIS analysis. At this stage, various types of spatial relationships between objects are analyzed, and some of the most popular relationships include the following:

    ° **point-to-point relationships**:

    "is within": All schools that are within a distance of 1 km from a living place

    "is nearest to": The primary school that is closest to a living place

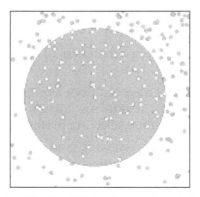

Example of the "is within" point-to-point relationship

° **point-to-line relationships**:

"is nearest to": The street that is closest to a subway station entrance

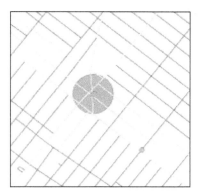

Example of the "nearest to" point-to-line relationship

° **point-to-polygon relationships**:

"is contained in": All public schools within a certain community boundary

Example of the "is contained in" point-to-polygon relationship

° **line-to-line relationships**:

"crosses": Whether a particular trail crosses a road

"is within": Find all trails that are within a distance of 1 km from a particular river

"is connected to": Find all streets connected to a particular highway

- ○ **line-to-polygon relationships**:

  "intersects": Find all districts crossed by a cycleway

  "contains": Find all streets that are completely within a certain district

Example of the "intersects" line-to-polygon relationship

- ○ **polygon-to-polygon relationships**:

  "completely within": Find all districts that are completely within a hazard zone

  "is nearest to": Find the building that is nearest to a park

Example of the "completely within" polygon-to-polygon relationship

4. **Primarily analyze and prepare the data**: The data is analyzed according to the set of criteria defined in the previous stages. Common analysis operations involve selection by location, buffering, rasterization, proximity (raster distance), and so on. After all the necessary layers are ready, they should be prepared for overlay, which involves reprojection into a common coordinate reference system (if necessary), setting ranges for various ranks, and reclassification in the common ranking system. All the layers should contain values in uniform units, otherwise their overlay will be meaningless and hard to interpret.

5. **Overlay the data and interpret the results**: Previously prepared layers are combined into a single coverage based on a set of user-defined rules. Depending on the data available and the rules applied, the following suitability assessments are possible:

   ° **Binary suitability assessment**: All of the area is divided into appropriate and inappropriate categories. This is the simplest type of assessment that can be obtained from vector data overlay.

   ° **Ranked suitability assessment**: Places are ranked from least appropriate to most appropriate based on the entire range of predefined criteria. This type of assessment can be derived from both vector and raster data. It lets you avoid simple yes/no assessments, which are not inherent to the real word. This advantage is counterbalanced by the subjectivity of data rankings and equal importance of various factors. Nevertheless, in the real world, their contribution to overall assessment can vary.

   ° **Weighted suitability assessment**: This is similar to the previous type of assessment and has only one significant difference: various factors can be weighted differently according to their importance for a certain type of activity. This type of assessments relies on the raster algebra approach and is thought to be all-inclusive, but not without some subjectivity, especially when it comes to factor weighting and interpreting the final result.

In the following sections, we will go through all of these stages individually and perform a weighted suitability assessment for living, combining various data types from the training dataset.

Step 1 – define the goal and objectives of our analysis

Throughout this tutorial, we will suppose that we are working on behalf of a young couple with a little child. They are looking for a perfect place to live in within a certain region of their interest. Our goal is to use the power of GIS-based suitability analysis methods and provide an objective and reliable answer to their question.

The goal of this analysis is to find areas that would be a good match for a young family with a child, given certain considerations. Many of their requirements are similar to those of a traditional housing estate development company. For example, proximity of subway stations, green zones, and public safety must be all taken into consideration. There are also some family-specific requirements that should be considered. As already mentioned, the family has a little baby, which means that we should take into account the existence of early childhood or elementary schools nearby. Also, they are keen on sports, and it would be great if the area they are going to live in has well-developed and active rest infrastructure. After this kind of review, we can formulate some more specific requirements and objectives:

- **Safety**: The area should not be exposed to various natural and crime hazards
- **Connectivity**: It should be well-connected to the city's transport network
- **Greenness and openness**: It should be close to parks or other green areas
- **Educational potential**: Early childhood or elementary schools should be located in the neighborhood
- **Active rest opportunities**: These include a cycling network and athletic facilities
- **Cultural life**: Art galleries and museums can be recognized as general signs of cultural pulse beating

Now that the main objectives and requirements have been clarified, we can proceed to the next step and study all of the available data to assess its relevance to our example.

Step 2 – analyze the available data and define its relevance

As soon as we have defined the basic requirements, we need to explore the data layers that are potentially useful and relevant for our analysis. The training dataset contains a large number of datasets, and the most relevant among them are listed in the following table:

Requirement	Layer	Potential relevance
Safety	hurricane_evacuation_zones	Hurricane evacuation zones are the areas of the city that may need to be evacuated due to life- and safety-related threats from a hurricane storm surge.
	hurricane_inundation_zones	Hurricane inundation zones are the areas of worst-case storm surge inundation.
	noise_heatmap	The raster created in the *Creating heat maps with the Heatmap plugin* section of *Chapter 5, Answering Questions with Density Analysis*, that shows the spatial density of registered noise complaints might be useful for potential assessment for public safety.
Connectivity	subway_entrances	Locations of subway entrances.
Greenness and openness	parks	A layer containing open space features, such as courts, tracks, parks, and so on.
	tree_density	This is a raster layer created from the tree census data.
Educational potential	elementary_schools	These are the point locations of schools based on the official addresses. This layer includes some basic information about the school, such as the name, address, type, and principal's contact information.

Requirement	Layer	Potential relevance
Active rest opportunities	`bike_routes`	Locations of bike lanes and routes throughout the city.
	`athletic_facilities`	This layer contains athletic facilities and some basic information about them, including primary sport type, surface, dimensions, and so on.
Cultural life	`museumart`	Locations of museums and art galleries.

These layers are primarily defined as most suitable for our analysis because they are directly related to the specified requirements, but most of them are represented as simple geometric primitives or density raster coverages that are difficult to be directly interpreted as suitability criteria. in the next stage, we will explore their main properties, including attributes, and will try to formulate specific and measurable criteria based on various types of spatial relationships between objects.

Step 3 – define the criteria of analysis

According to the description of `hurricane_evacuation_zones`, there are six zones, ranked in the attribute field zone by risk of storm surge impact, with zone 1 being the region most likely to be flooded. In the event of a hurricane or tropical storm, residents in these zones may be ordered to evacuate. Areas with a zone value of x are not in any evacuation zone. Areas with a zone value of 0 are any of the following: water, small piers, or uninhabited islands. For the purpose of analysis, this layer should be rasterized and ranked according to the risk of hurricane impact, with rank values descending from areas that are not in any evacuation zone to those most likely to be flooded.

The `hurricane_inundation_zones` polygon layer contains information about the risk of storm surge inundation in the `category` attribute field, in which the value is the surge height in feet. Areas that are most likely to be inundated are assigned a value of 1, and areas that are excluded from inundation modeling are assigned a value of 5. This layer should be rasterized and ranked with the highest potential suitability values for excluded areas, and the lowest for the areas of the 1 category.

The `noise_heatmap` raster layer is a raster that should be ranked using several categories, with the lowest suitability values for the noisiest places and vice versa. A good thing here is that we don't need to rasterize this layer, as we did with the previous layers. At the same time, establishing the amount and ranges for the rankings brings subjectivity into our assessment. The `tree_density` layer, which is also a density raster, should be analyzed similarly.

The other selected layers should be analyzed first for their proximity. For this purpose, we will first rasterize them, then create continuous proximity rasters, and finally rank them under several categories according to the proximity values (the closer an object, the higher the suitability value). Again, in the case of user rankings, we will not be able to avoid some subjectivity in our assessments. Also, the final proximity rasters can be weighted according to their importance in the overall suitability assessment.

Step 4 – Analyze and prepare the data

There are three main approaches to primary data analysis. These depend on the initial data type and available attributes:

- **Rasterizing and ranking categorized vector layers**: These are the layers that already contain all the necessary values, and at the preparation stage, all of them should be rasterized to the similar extent and resolution. Also, their categories should be ranked properly, with the highest values for the most suitable areas and vice versa. Examples of these layers are `hurricane_evacuation_zones`, `hurricane_inundation_zones`, and so on.

- **Ranking density rasters**: These are raster heat maps that should be converted from continuous coverages to categorized values where the highest value symbolizes the most appropriate area, and the lowest is related to the least suitable area. Examples of these layers are `noise_heatmap` and `tree_density`.

- **Generating and ranking proximity rasters**: This is the most tedious workflow. Vector layers should be rasterized first, and then proximity rasters should be created and ranked properly. This category encompasses the following vector layers: `subway_entrances`, `parks`, `public_schools`, `bike_routes`, `athletic_facilities`, and `museumart`.

Note that for the final output, we will always have ranked unique value rasters, with the highest values denoting the most suitable areas. Also, it is important that all output rasters share a common extent, which is necessary for their proper overlay with the raster calculator and overall suitability assessment.

In the upcoming sections, we will go through the previously mentioned workflows for the example of raster layers. As soon as you grasp the principle, you will be able to prepare other layers independently.

Rasterizing and ranking categorized vector layers

In this example, we will work on the `hurricane_evacuation_zones` layer. The attribute we are particularly interested in is `zone`. This is because by this attribute, areas are prioritized for evacuation. Areas with the lowest value, 1, are the most likely to be evacuated, and vice versa. In this case, we can use these values directly to rank the raster. We have already done rasterization in *Chapter 6, Answering Questions with Visibility Analysis*, in the *Step 1 – converting a buildings' vector layer to raster* section. This time, we will follow the same procedure.

One thing that prevents us from doing rasterization directly is that the attribute field that is used for rasterization should be numeric. If you check out the attribute field in the **Fields** section under **Properties**, you will see that the field zone has **Type QString**, and its values that contain numbers 1-6 and letters (X) are interpreted not as numbers but as sequences of symbols or strings. That's why this field is unavailable for rasterization and should first be converted to numbers. This can be done easily with **Field calculator**:

1. Open the attribute table of the layer using the right-click **Open Attribute Table** shortcut, or hit the relative button 🖽 from the **Attributes** toolbar. In the attribute table toolbar, either click on the **Open field calculator** button or use the *Ctrl + I* keyboard shortcut.

2. First of all, we should get rid of the X value that cannot be interpreted as a number and cannot be converted to it. As we have only one line that contains the X value, we simply toggle editing mode by clicking on the ✏ button in the attribute table toolbar. Double-click on the cell and manually enter the new value of 7. If you have multiple values to change, you can use the following expression in **Field calculator** to change some zone field values:
   `CASE WHEN "zone" = 'X' THEN '7' ELSE "zone" END`.

3. Click on the 🧮 button to open the **Field calculator** dialog window as shown in the following screenshot, and make the following adjustments:

   ° Make sure that the **Create a new field** toggle is on.

   ° Type the **Output field** name manually, for example, `rank`.

   ° Select **Whole number (integer)** from the **Output field** type list, as we are going to use short integers for ranking.

   ° Reduce **Output field width** to 1. This is because the rank values don't exceed 10 and we don't want to create excess data by producing fields of length that exceed actual data values.

4. In the **Expression** field, we need to type a function that will be used to create the values of a new field. In the **Functions** list, expand the **Conversions** item and double-click on the **toint** function. According to its description, it converts a string to an integer number. Nothing changes if a value cannot be converted to integer (for example, `123asd` is invalid). After double-clicking, the function will be added to the expression with an open bracket, after which you should type (or double-click to add an item from **Fields and Values**) the field name to be converted in double quotation marks, and close the brackets. In our case, the resulting expression will be `toint( "zone" )`.

5. After you've clicked on the **OK** button, the new field will appear at the end of the table. Deactivate editing mode, confirm saving of the edits, and exit the attribute table window.

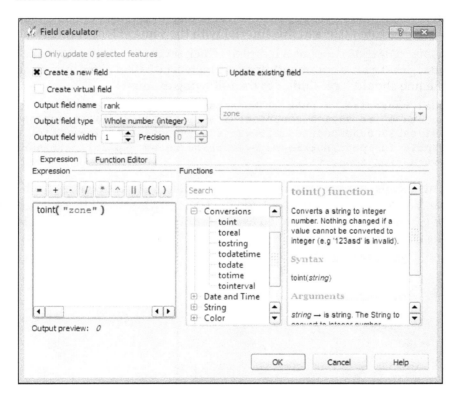

The layer is ready for rasterization. Open a dialog window by going to **Raster** | **Conversion** | **Rasterize (Vector to Raster)**. In this dialog window as shown in the following screenshot, adjust the following settings:

1. From the **Input file (shapefile)** drop-down list, select `hurricane_evacuation_zones`.

2. From the **Attribute** drop-down list, select `rank`.

3. In **Output file for rasterized vectors (raster)**, click on the **Select** button. Navigate to your working directory, and type `hurricane_evacuation_zones.tif` as the new layer name. This message will be shown: **The output file doesn't exist. You must set up the output size or resolution to create it.**. Click on **OK** and proceed to the next stage.

4. In the previous steps, you defined some major options. For convenience, we will create rasters in this tutorial with the same extent and resolution using `lidar_dem.tif` as a template. Click on the ⟋ button to make the `gdal_rasterize` command-line parameters editable. After modification, the line should look similar to the following example:

```
gdal_rasterize -a rank -l hurricane_evacuation_zones -a_
nodata 0 -te 982199.3000000000465661 188224.6749999999883585
991709.3000000000465661 196484.6749999999883585 -tr 10 10 -ot
UInt16 fullpath/hurricane_evacuation_zones.shp fullpath/hurricane_
evacuation_zones.tif
```

This means that the output raster will contain rasterized values from the rank attribute field. Areas outside the layer polygons will be assigned a value of 0, which will be interpreted as `nodata`. The output data type is a 16-bit unsigned integer.

5. Click on the **OK** button. The result of rasterization will appear in the **Layers** panel.

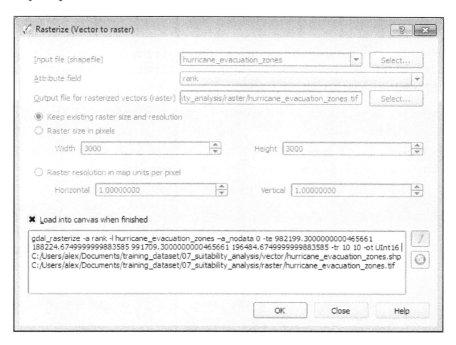

The `hurricane_inundation_zones` layer should be preprocessed in a similar way. The layer contains some `zone` numbers. These can be interpreted as the severity of inundation risk; that is, the higher the number, the lower the risk. This means that we can use these values directly for ranking. In the case of this layer, the `gdal_rasterize` command-line parameters will look as follows:

```
gdal_rasterize -a category -l hurricane_inundation_zones -a_nodata 0 -te
982199.3000000000465661 188224.6749999999883585 991709.3000000000465661
196484.6749999999883585 -tr 10 10 -ot UInt16 fullpath/hurricane_
inundation_zones.shp fullpath/hurricane_inundation_zones.tif
```

When combined, these layers can give a cumulative assessment of suitability based on risk's severity from natural hazards.

Ranking density rasters

In the case of density rasters, we can use the ready `noise_heatamp.tif` created in the *Creating heat maps with the Heatmap plugin* section of *Chapter 5, Answering Questions with Density Analysis*, and create the `tree_density.tif` layer ourselves. First, we will prepare a `noise_heatmap.tif` that is originally much larger by extent than the area of interest that we are working on in this chapter. To clip the raster layer go to **Raster | Extraction | Clipper**:

1. From the **Input file (raster)** drop-down list select **noise-heatmap**, which is the raster to be clipped.

2. In **Output file**, click on the **Select** button to set a path and name for the output file, for example, `noise_heatmap_clip.tif`.

3. In the **Clipping mode** section, there are two possible modes:

   ° **Extent** is where you can enter the bounding box's coordinates manually or by dragging the map canvas. Use this approach to set the extent for your output file. Just remember that the extent should exceed the boundaries of the area of interest set by the `zipcode_bound` shapefile.

   ° If **Mask layer** is active, you can select a polygonal shapefile, and it will be used as a clipping boundary.

   After you have clicked on the **OK** button, the layer will be loaded into the map canvas.

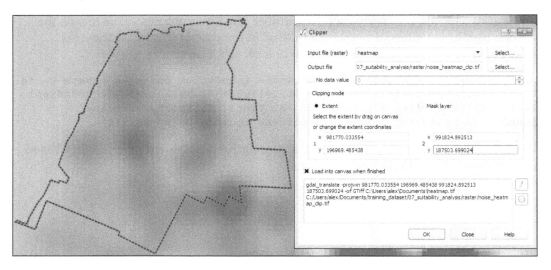

The following steps should be done to assess its value range, divide it into categories, and rank them:

1. Double-click on the `noise_heatmap_clip` layer to open the **Layer Properties** window. In this window, go to the **Metadata** section. Explore the **Properties** window in the lower part of the dialog until you find the **Band 1** highlighted line with basic raster statistics, as shown here:

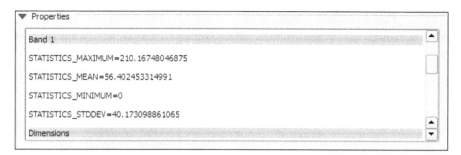

2. We will deal with the density raster, and its values are interpreted as the number of noise complaints per pixel area. This continuous surface should be divided into categories and they should be ranked according to the number of complaints. For this, a certain critical number of complaints should be set. There are no officially recognized limits and requirements, but we can interpret the mean dataset value as some critical number (it is slightly greater than 56, but we will use 50 for convenience). Divide the values into categories and assign them the following ranks (the less the number of complaints, the better):

Category	Value range	Suitability rank
1	Less than 50	5
2	50 to 100	4
3	100 to 150	3
4	150 to 200	2
5	Greater than 200	1

3. For the rankings, go to **Raster | Raster Calculator** and adjust the following options in its dialog window:

    ○ Enter the **Output layer** path and name, for example, `noise_ranked`

    ○ Select the `noise_heatmap_clip @1` raster band and click on the **Current layer extent** button to make sure that the output layer will have the same resolution and extent

○ In the **Raster Calculator** expression window, enter the following expression:

```
("noise_heatmap_clip@1" <= 50) *5 + ("noise_heatmap_clip@1"
> 50 AND"noise_heatmap_clip@1" <= 100) *4 + ("noise_heatmap_
clip@1" > 100 AND "noise_heatmap_clip@1" <= 150) *3 +
("noise_heatmap_clip@1" > 150 AND "noise_heatmap_clip@1" <=
200) *2+ ("noise_heatmap_clip@1" > 200)*1
```

This expression means that every pixel that falls under a certain range given in brackets first assigned 1 value, and then it is ranked by a certain multiplier value, which is outside the brackets. After you click on the **OK** button, the reclassified raster will be loaded into the **Layers** panel.

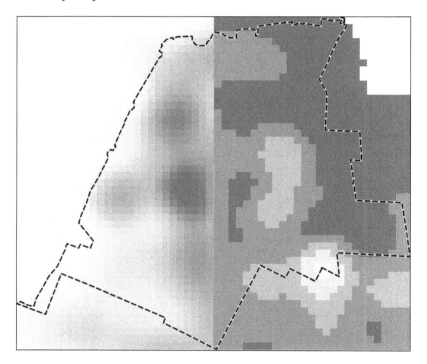

Pay attention to this: due to ranking, the heat map was inverted; that is, the noisiest hotspots get the lowest ranks and the quietest places get the highest.

In a similar way, we can create and rank heat maps for other point objects that are of interest in our suitability analysis, namely `trees` and `museumart`.

Generating and ranking proximity rasters

The workflow will be explained in the example of the `subway_entrances` point vector layer:

1. First, the layer should be rasterized in a common way. The `gdal_rasterize` line will contain the following parameters:

   ```
   gdal_rasterize -l subway_entrances -burn 1 -a_nodata
   0 -te 982199.3000000000465661 188224.6749999999883585
   991709.3000000000465661 196484.6749999999883585 -tr 10 10 -ot Byte
   fullpath/subway_entrances.shp fullpath/subway_entances.tif
   ```

   In the output layer, existing point locations will be marked by pixel values of 1, while all other areas will be assigned the `nodata` values of 0.

2. Go to **Raster** | **Analysis** | **Proximity (Raster Distance)**. This dialog generates a raster proximity map that indicates the distance from the center of each pixel to the center of the nearest pixel identified as a target pixel. Target pixels are those pixels in the source raster for which the raster pixel value is in the set of target pixel values. If not specified, all nonzero pixels will be considered target pixels. In the dialog window, adjust the following parameters:

   ○ From the **Input file** drop-down list, select the `subway_entrances` layer.

   ○ In the **Output file**, click on the **Select** button and specify a path and a name for the output layer, for example, `subway_entances_proximity.tif`.

   ○ Make sure that the **Dist units** parameter is activated and set to **GEO**. In this case, the distances generated will be in georeferenced coordinates (feet).

The resulting parameters line will look like this:

```
gdal_proximity.bat fullpath/subway_entances.tif Ifullpath/subway_
entances_proximity.tif -distunits GEO -of GTiff
```

To convert a vector to raster, we used a tool from the GDAL. It is called `gdal_proximity`. You can read an extensive synopsis of the `gdal_proximity` parameters and their values at `http://www.gdal.org/gdal_proximity.html`.

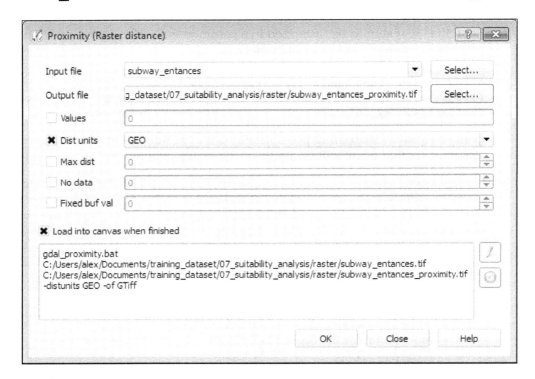

3. The proximity raster should be divided into discrete categories and they should be ranked. For this operation, we will use **Raster Calculator**. The major problem here is to decide on the amount and properly select the proximity categories for ranking. The optimal decision depends on the so-called walking radius, that is, the distance that people are comfortable to walk up to. Generally, transport planners have observed that the walking distance that most people seem to cover comfortably — beyond which ridership falls drastically — is about 400 m (approximately 1300 feet). We will apply this 400 m rule to categorize the layer values (min as zero and max as 4988.71) into four categories, with the following ranks:

Category	Proximity values range (feet)	Suitability rank
1	Less than 1,300	4
2	1,300 to 2,600	3
3	2,600 to 3,900	2
4	Greater than 3,900	1

Go to **Raster | Raster Calculator** and adjust the main options. Enter the following expression to generate a new `subway_entrances_proximity_ranks` raster:

```
("subway_entances_proximity@1" <= 1300) *4 +
("subway_entances_proximity@1" > 1300 AND
"subway_entances_proximity@1" <= 2600) *3 +
("subway_entances_proximity@1" > 2600 AND
"subway_entances_proximity@1" <= 3900) *2 +
("subway_entances_proximity@1" > 3900) *1
```

In the following screenshot, you can see what continuous (on the left) and ranked (on the right) rasters may look like:

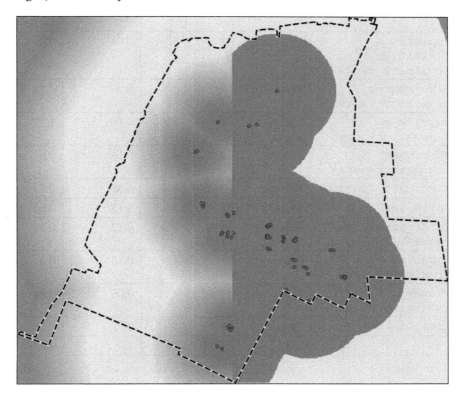

In a similar way, you can preprocess other vector layers that should be analyzed from the proximity position, namely `parks`, `bike_routes`, `athletic_facilities`, and `elementary schools`. As a result, you will have a set of raster layers ranked with several categories according to the proximity of selected objects. The higher the rank, the closer the objects. As a general rule of thumb, apply the 400 m (or 1300 feet) value to rank the rasters properly.

Step 5 – overlay the data and interpret the results

Now we have everything ready for overlaying the rasters and generating a cumulative suitability assessment. In the following table, you can see the full list of layers, weighted by their importance for general suitability. Weights are simple coefficients within a range of 0 to 1, and they are used to modify the rank properly.

No.	Raster layer name	Raster layer type	Rank min	Rank max	Weight coefficient	Max possible value = max rank * weight
1	athletic_ facilities_ proximity_ranks	Proximity	1	5	0.04	0.2
2	bike_routes_ proximity_ranks	Proximity	1	3	0.06	0.18
3	hurricane_ evacuation_zones	Zoning	1	7	0.16	1.12
4	hurricane_ inundation_zones	Zoning	1	5	0.16	0.8
5	Museumart_ranked	Density	1	4	0.05	0.2
6	Noise_ranked	Density	1	5	0.13	0.65
7	Parks_proximity_ ranks	Proximity	1	3	0.12	0.36
8	Schools_ proximity_ranks	Proximity	1	4	0.15	0.6
9	subway_entrances_ proximity_ranks	Proximity	1	4	0.06	0.24
10	Tree_ranked	Density	1	4	0.07	0.28
Total			10	44	1.0	4.63

The expression to be used for our assessment of suitability can be constructed by several steps:

1. All the available factors that are represented by raster layers should be multiplied by their weight coefficients and summarized, like this: `(factor_1*weight + factor_2*weight + factor_3*weight + … factor_n*weight)`. This kind of assessment gives gross suitability, which is difficult for interpretation because the obtained values are not calculated relatively to the possible minimum and maximum.

2. For simplicity of interpretation, a gross suitability assessment can be divided into the sum of the maximum possible value. The extended formula will look like this: *(factor_1*weight + factor_2*weight + factor_3*weight + … factor_n*weight) / (factor_1_max*weight + factor_2_max*weight + factor_3_max*weight + … factor_n_max*weight)*. As a result, the output value range will be from 0 to 1, where the maximum suitability values are close to 1.

3. Optionally, the result can be multiplied by 100. Then the output values will be a percentage of suitability.

Go to **Raster | Raster Calculator** to perform an assessment:

*   Set the path and name for **Output layer**

*   Select a layer from the **Raster bands** list to set up **Current layer extent**, for example, `hurricane_inundataion`

*   In the **Expression** window, enter the following formula (if you are not sure about the values, check out the preceding table):

```
( ( "athletic_facilities_proximity_ranks@1" * 0.04 +
"bike_routes_proximity_ranks@1" *
0.06+"hurricane_evacuation_zones@1" *
0.16+"hurricane_inundation_zones@1" *
0.16+"museumart_ranked@1" * 0.05+"noise_ranked@1" *
0.13+"parks_proximity@1" * 0.12+"schools_proximity@1" *
0.15+"subway_entrances_proximity_ranks@1" *
0.06+"tree_ranked@1"*0.07) / 4.63 ) *1000
```

After you have clicked on the **OK** button, the resulting layer will be added to the map canvas. The suitability raster value's range varies from 42 to 96 percent. Thus, it can easily be classified and interpreted. Navigate to **Layer properties | Style** and adjust the rendering properties to those shown in the following screenshot:

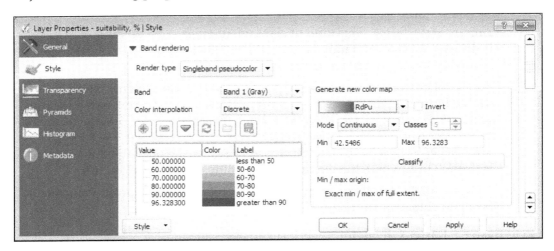

After applying these settings, the layer will look as follows:

We can use this raster as a basis for a primary visual suitability assessment of the area of our interest, or go further and combine it with other layers to identify the exact blocks that match best the entire range of suitability criteria.

In that case, we need to perform the inverse sequence of steps: select the most suitable areas, vectorize, and overlay the polygons of maximum suitability with residential living areas to identify the blocks and buildings that would be particularly suitable:

1. The initial suitability raster should be categorized into only two classes by a limit of 90 percent suitability. Go to **Raster | Raster Calculator** and define the path and name for the output raster (for example `max_suitability`). In the **Expression** window, enter `"suitability, %@1" >= 90`. On the next page, you can see that the resulting layer contains only two classes: suitable (the value is 1; assigned to areas with suitability greater than or equal to 90 percent), and unsuitable (the value is 0).

2. Now we need to vectorize these areas to be able to query vector layers with them. Open the **Polygonize (Raster to vector)** dialog window by going to **Raster | Conversion** and adjust the following parameters:

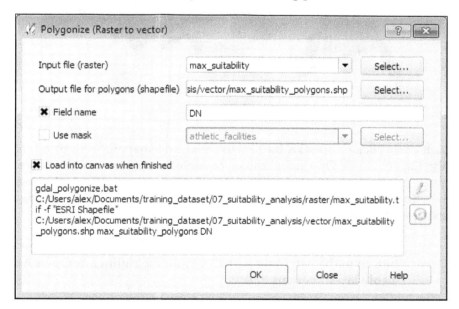

-   Select `max_suitability` as the raster to be polygonized from **Input file (raster)**.

- ° Provide the path and name of the output shapefile in **Output file for polygons (shapefile)**, for example, `max_suitability_polygons`.
- ° Activate the **Field name** toggle and accept the default DN value. This option is responsible for creating and populating field with class values from the initial raster.

3. After you click on the **OK** button, the resulting shapefile will be added to the **Layers** panel. Originally, the layer contains polygons of all classes, whereas we are interested only in class 1. For the purpose of selecting and deleting unnecessary polygons, we will use the **Select features using an expression** option:

- ° Open the `max_suitability` attribute table by clicking on the ⊞ button in the **Attribute** panel, or from the right-click **Open Attribute Table** layer shortcut.

- In the attribute table toolbar panel, click on the **Select features using an expression** button, ε , and enter the following expression: `"DN"` `= 0`. After clicking on **OK**, all the polygons that satisfy the condition will be highlighted in the attribute table and on the map canvas.

- As we don't need these polygons for further analysis, we should delete them. In the attribute table toolbar panel, click on the ✏ button to toggle editing mode, or use the *Ctrl + E* keyboard shortcut. Now we can execute **Delete selected features** using the 🗑 button, or just press *Del* from the keyboard.

- After deleting the unnecessary data, don't forget to save your edits (using 💾 or *Ctrl + S*) and deactivate editing mode. In the following screenshot, you can see that the layer contains only those polygons that encompass the maximum suitability values, which are set to 90 percent:

4. Now this layer can be used to overlay with other vector layers and analyze spatial relationships between objects, as was described in the *Basics of suitability analysis* section. For example, we can identify the zoning areas of primary residential living area that are of potential interest to us according to the suitability criteria:

   ○ Open a dialog window by going to **Vector | Research tools | Select by location**. In this dialog, you should first select the layer from which objects are to be selected — residential_zoning from the **Select features in:** drop-down list — like this:

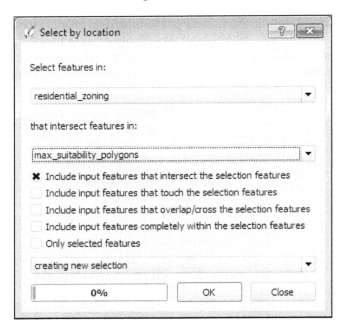

   ○ the **that intersect features in:** drop-down list, select max_suitability_polygons, which will be used as a selector.

   ○ There are several overlay selection options. Activate **Include input features that intersect the selection features**. Only those polygons that are within the query mask boundary or intersect it will be added to the selection. Click on the **OK** button. You will see the following result on the map canvas:

5. Similarly, you can try another type of query and identify the exact buildings that satisfy a suitability condition. In that case, the dialog window will look as follows:

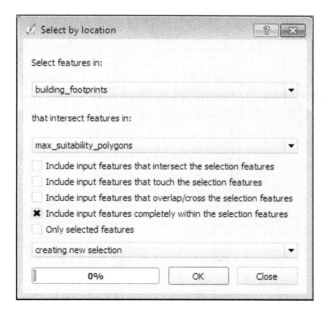

After clicking on the **OK** button, you will see the following results on the map canvas:

Notice that this time, we have selected buildings that were completely within the maximum suitability areas. So, we are ready to provide a solid and specific answer to the question, "Which is the best place to live?"

Summary

In this chapter, you followed the entire path of spatial analysis, from raising questions to providing concrete answers, that is, finding a specific place and objects on a map. You explored some basics of suitability analysis and saw how it defines and interprets spatial relations between objects. Also, you explored a lot of operations related to data preprocessing: rasterization, reclassification, raster algebra performance, and so on. Finally, you reviewed spatial queries that are based on overlaying spatial objects and analyzing spatial relationships between them.

Executing analysis operations step by step is good when you learn, but if you have a lot of data and limited time to provide the answer, then you should improve the performance of your analysis. This can be done with the QGIS Processing framework. It provides access to hundreds of geoprocessing tools (from the simplest to the most sophisticated).

In the next chapter, we will introduce QGIS Processing, and you will learn how to automate spatial analysis routines by creating your own chains of instruments (models).

8

Automating Analysis with Processing Models

In this chapter, you will learn how to use the graphical modeler from the QGIS Processing framework to create models and automate complex analysis tasks. We will cover all the necessary steps starting from creating the model and ending up with sharing the model with other users.

In this chapter, we will go through the following topics:

- The QGIS Processing framework
- The Graphical modeler
- Adding inputs
- Implementing the workflow
- Filling in model metadata and saving
- Editing models
- Sharing models

The QGIS Processing framework

The QGIS Processing framework (formerly SEXTANTE) is an QGIS core plugin that implements a powerful analysis and geoprocessing environment and provides a user-friendly interface for various native QGIS algorithms and third-party analysis software, such as GRASS, SAGA, Orfeo ToolBox, and many others.

Refer to the QGIS user guide at `http://docs.qgis.org/ testing/en/docs/user_manual/processing/intro. html` if you want more information about the QGIS Processing framework.

The Processing user interface consists of four main components. Each of them allows you to run algorithms in a different way. Deciding which component to use depends on the task and type of analysis. As the Processing plugin is activated by default, we can access all of its components and functionality, except the batch processing interface, from the Processing menu:

- Toolbox: This is the main Processing GUI element. It provides access to all available algorithms (grouped by provider or purpose) and allows us to run them in single-pass and batch modes. The Processing Toolbox has two modes: simplified (default) and advanced. In simplified mode, all algorithms are placed in predefined groups and have user-friendly names, with the aim of helping novice users find the necessary tools. In advanced mode, algorithms are grouped by *providers*. Each provider represents some analysis package or program, for example, GRASS, Orfeo ToolBox, and so on. Also, in advanced mode, we have access to a larger number of algorithms, as some special providers are available only in this mode.

  In the upcoming sections, we will assume that the toolbox is used in advanced mode.

  To activate advanced mode, just select the **Advanced** interface from the combobox at the bottom of the Processing toolbox.

- **history manager**: This stores information about all executed algorithms and their parameters so that, if necessary, one can easily reproduce a past action. Also, all errors, warnings, and information messages are displayed here.

- **batch processing interface**: This is used to run individual algorithms on multiple datasets. This interface is accessible only from the Processing Toolbox and is not available in the Processing menu.

- **graphical modeler**: This is used to create new algorithms by combining existing ones into a single workflow. It allows us to easily automate a complex analysis that involves several steps.

In the next section, we will see how to use the Processing Graphical modeler and create our own models to automate analysis tasks.

Graphical Modeler

As we have already seen in previous chapters, even a simple analysis may require several steps. In real-life applications, much more complex tasks exist, and they involve and combine different processes of analysis. Running them manually is a time-consuming process, and it may take hours to complete. Things become even worse and complicated when you need to run this analysis several times with different input data or use the same data with different settings.

This is where the QGIS Processing framework comes to help. It allows us to automate repeated tasks and create new complex algorithms that involve different types of data processing. With the help of the Graphical modeler from the Processing toolbox, we can easily create our own algorithms by combining existing algorithms. The created model can be executed like any other Processing algorithm, and even used as part of another, more complex model.

There are two common ways of opening the Graphical Modeler:

- From the menu, go to **Processing | Graphical Modeler...**.
- Use the Processing Toolbox. Ensure that the Processing Toolbox uses the **Advanced** mode. Go to the **Models** group and expand it. In the **Tools** subgroup, find the **Create new model** item and double-click on it.

Any of these methods will open the **Processing modeler** window, as shown in the following screenshot:

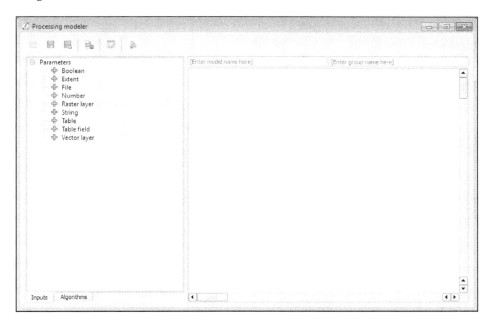

The **Processing modeler** window can be divided into the following main areas:

- **Toolbar**: This provides some useful actions, such as opening existing models, saving and exporting a model, the help editor, and so on.

- **Inputs and algorithm trees**: These are located on the left side of the window and consist of two tabs: one contains the available inputs, and the second contains the available algorithms. It is worth mentioning that the **Algorithm** tab supports the same modes for displays of algorithms as the toolbox: simplified and advanced. However, note that there are no switches in this tab for changing the display mode. If you need to change it, you should do it in the toolbox before opening the modeler.

- **Working area**: This is used to display the model structure. Here, we will place our building blocks and connect them to each other.

The creation of the model can be divided into two steps:

1. Defining the input: This means that we need to specify which data is necessary for analysis. All items of input that we define here will be represented later as algorithm parameters, and the user will be able to set them according to their requirements.

2. Defining the workflow: At this step, we establish links between input items and algorithms. In other words, we determine how each algorithm in the model will use input and output from other algorithms.

Let's create a model to generate density maps from a point vector layer using the binning technique, which was described in the *Mapping density with a hexagonal grid* section of *Chapter 5, Answering Questions with Density Analysis*. To make this example more useful, we will create a model that generates not only a hexagonal density grid, but also a square density grid. This is done so that you can later compare the results and choose the one that you like more.

Adding inputs

The first step of model creation is defining the input necessary for it. To do this, we should go over all the algorithms of the process that we want to automate and find out which input items are necessary. It is necessary to remember that some algorithms may not need separate inputs. They will use only results obtained in some previous steps, or in other words, the output of other algorithms. Also, some inputs may not be important and can be hardcoded in the model.

The **Processing modeler** supports the following kinds of input, which can be found in the **Inputs** tab on the left side of the modeler window:

- Number: This is used for integer and floating-point values. When adding it, it is necessary to specify the minimum and maximum values allowed as well as a default value.

- String: This is a string literal. There is only one additional setting, namely the default value.

- Boolean: This is a boolean value, usually used as a checkbox. It is necessary to specify the default state: checked or unchecked.

- Extent: This represents a geographical extent.

- Vector layer: This is used for vector layers. If necessary, it is possible to limit the supported geometry types to one of the available types and make this input optional.

- Raster layer: This represents the raster layer in GDAL-supported format. It can be optional.

- Table: This is used for geometryless tables, for example, DBF files. It can be optional.

- Table field: This represents a field of the layer attribute table or geometryless table. It is necessary to specify the parent layer from which this field will be fetched.

- File: This is used to represent files and directories.

To add an input, double-click on its name. A **Parameter definition** dialog will open. Its content depends on the input type, but common for all input types is a **Parameter name** field, where it is necessary to specify at least the name of the input. This text will be used as a caption for the corresponding field when the model will be executed. All other fields are different for different input.

 You can also use the drag-and-drop method from the inputs/ algorithms tree to the modeler work area when adding inputs or algorithms.

Once all the settings are specified, click on the **OK** button. A new input block will be added to the modeler's working area.

Let's start with adding input for our model. The first input is obvious; it is a vector layer to create a density map for. Double-click on the **Vector layer** item in the **Inputs** tab of the **Processing modeler** window to open the **Parameter definition** dialog, which looks like this:

Enter a name in **Parameter name**, for example, Point layer. As we don't want to see all available layers as input later, and search for a required layer through a long list, we can limit the supported geometry type by choosing the **Point** geometry type in the **Shape type** combobox. Finally, as this input is required, we leave the **Required** field as **Yes**. Click on the **OK** button to add the input to the modeler working area.

To be able to make our final density map more detailed or more general, depending on our requirements, we need a way to adjust the cell size. So, double-click on the **Number** item in the **Inputs** tab to add another input.

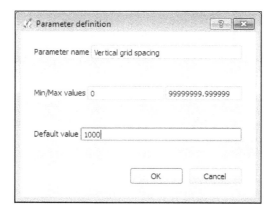

Enter Vertical grid spacing as **Parameter name** as shown in the preceding screenshot. We can leave all other fields untouched, but it is better to specify them too. This should be done to prevent wrong input from users. We should now keep in mind the fact that different layers may have different coordinate reference systems (CRS). As a result, we will probably have different distance units. So, our numeric input should allow the user to enter values suitable for layers in geographic CRS (degrees) and projected CRS (meters, feet, and so on). That's why we use 0 as the minimum value and 99999999.999999 as the maximum value. These values allow us to change the cell size over a wide range, irrespective of the CRS used. Enter any reasonable value as **Default value** and click on **OK** when you're done.

As cell sizes in the horizontal and vertical directions can be different, we need to create another numeric parameter called Horizontal grid spacing using the same settings as you saw before.

> If you always need the same spacing in both the horizontal and vertical directions, it is better to use only one input parameter, as a single parameter can be used multiple times as an input in the same algorithm, or even in different algorithms. Later, we can adjust our model to use only one numeric parameter to define the grid spacing.

As you may remember from the *Mapping density with a hexagonal grid* section of *Chapter 5, Answering Questions with Density Analysis,* when using a Create grid algorithm, it is necessary to specify the desired grid extent, and we take this extent from our input layer.

Now, for designing our model, we have two options:

1. Define the input for the grid extent. This will make our model a bit more complex, as we need to specify the grid extent each time we use the model, instead of specifying just the layer. But this option also brings more flexibility, because by having a separate input for the grid extent, we will be able to generate density maps not only for the entire layer but also for a specified region inside it.

2. Take the extent from the input layer. In this case, our model will be simpler for users, as all they need to do is to define the layer and grid cell size. But on the other hand, the density map will be generated for the entire layer.

As the most common use case is a density map for the entire layer, we choose the second option. The grid extent should be calculated automatically from the input point layer.

So, we have just defined almost all of the necessary input for our model: the input layer, and two numbers that represent the grid cell size in the horizontal and vertical directions. There is only one undefined input—the grid extent. You will learn how to extract it from the input layer in the next section, as this is closely related to workflow definition.

Implementing the workflow

When all the input items are in place, we can start implementing a workflow. All the available algorithms can be found in the **Algorithms** tab on the left side of the modeler window.

Workflow implementation is very similar to manual execution of all steps of analysis: we add algorithms one by one, choosing the correct input and, if necessary, defining the output. While the modeler allows us to add parameters and algorithms in any random order, it is better to add algorithms in the same order in which they should be executed to avoid confusion. Let's start!

As you may remember, we did not add a separate input for grid extent, and you may be curious to know how we will extract it from the point layer in a format compatible with the modeler input. The answer is simple: we will use special tools, which are so-called modeler-only tools. What are these? Well, all the algorithms that are available exclusively in the **Processing modeler** are not available in the toolbox. Basically, these are helper tools that can be really useful in different models, for example, a simple calculator or a value converter. You can find these tools in the **Algorithms** tab (as they are also algorithms) within the Modeler-only tools group.

Right now, we need a **Vector layer bounds** tool, which takes a vector layer as the input and returns its extent as the output.

 Besides the extent, this tool also returns four numeric values: minimum and maximum x coordinates, as well as minimum and maximum y coordinates. You may use these values as input for different algorithms.

This output value can then be used as the input in any algorithm that requires the extent as the input. Find the **Vector layer bounds** algorithm by typing its name in the filter field, and double-click on it to open the algorithm definition dialog as shown:

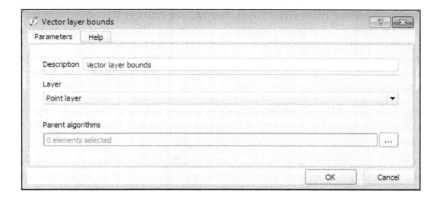

The dialog is very simple. We will discuss definition dialogs a bit later. Right now, we just select our previously added **Point layer** input in the **Layer** combobox and click on the **OK** button to close the dialog and add the algorithm to the modeler working area. Now that all preparation steps are done, we are ready to implement the main workflow.

The first step in our analysis is grid generation, so switch to the **Algorithms** tab, find the `Create grid` algorithm by typing its name in the filter field, and double-click on it to open the algorithm definition dialog.

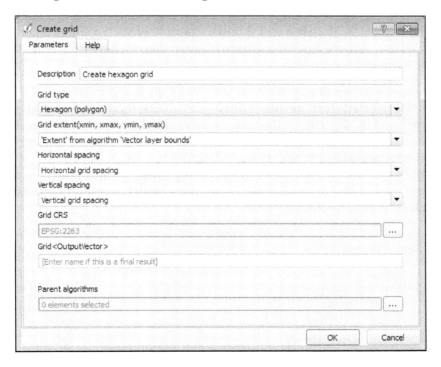

As you can see, the algorithm definition dialog has almost the same content as the algorithm dialog opened from the toolbox. The main differences are as follows:

- There is no **Log** tab where the algorithm reports its execution progress and other information. As the algorithm is not executed right now, this tab is not needed.

- There is a new **Description** field. It is used to define the algorithm name inside the model. By default, this name is the same as the original algorithm name. This feature is very handy when the model contains several algorithm blocks that represent the same algorithm but use different inputs. In such a case, just assign different descriptions to this algorithm, and you will never be lost amid them. You will be able to identify their outputs easily.

- There is a new **Parent algorithms** parameter at the bottom of the definition dialog, which does not exist when the algorithm is executed from the Processing toolbox. As the name implies, this parameter allows us to define parent algorithms for the selected algorithm, or in other words, set the algorithms' execution order. Parent algorithms will be executed in the specified order and before their children.

  By default, when the current algorithm uses the output of another algorithm as its input, the latter automatically becomes a parent of the current algorithm, and a link between these two algorithms is created. But sometimes, one algorithm can depend on another even if it has not used any output from it. Here is a well-known example of such a situation: before executing a query against a database layer, it is necessary to create the database and the layer.

- There is a different approach for selecting inputs and outputs. All values can be selected from the list of already available model inputs and outputs, generated by algorithms that are already added to the model algorithms. Also, values such as numbers, strings, booleans, and table fields can be entered manually in the corresponding fields. Note that you cannot change values that are entered manually at design time if your model is executed from the toolbox or used as part of another model. So, don't hardcode values unless there is an urgent need.

Now, proceed with the `Create grid` algorithm definition for our model. As we will generate two grids—hexagonal and square—using the same algorithm, we need to distinguish the algorithm blocks. So, change the **Description** field to `Create hexagon grid`. We hardcode the **Grid type** parameter by selecting `Hexagon (polygon)` from the combobox, as this is what we need.

In the **Grid extent** combobox, select `'Extent' from algorithm 'Vector layer bounds'`. This is the extent of our input layer, so the generated hexagonal grid will cover it.

> In the `'Extent' from algorithm 'vector layer bounds'` text, `'Extent'` is the name of the output from the algorithm, and `'Vector layer bounds'` is an algorithm description from the **Description** field of the corresponding definition dialog. So, you can easily detect which output is from which algorithm.

For horizontal and vertical grid spacing, select the previously defined input values from the corresponding comboboxes. You may notice that apart from the input values defined by us, there are some additional values produced by the **Vector layer bounds** algorithm. Ignore them.

Don't enter any text in the **Output** field, as the hexagonal grid generated by this algorithm is an intermediate, temporary output and not the final result. Temporary outputs will be generated and stored in the temporary directory, but will never be added to the QGIS main canvas after model execution. If you want to see these intermediate results, it is necessary to turn them into model output by entering descriptions for them.

When all parameters have their values assigned, click on the **OK** button to add the algorithm to the modeler's working area. You will see that the algorithm block is now connected to its inputs. Clicking on the small **+** sign near the **In** or **Out** labels will expand a list of inputs and/or outputs, so you can verify the correctness of the connections. The second click will collapse them back.

Add another **Create grid** algorithm. Change its **Description** field to `Create square grid`, select `Rectangle (polygon)` as **Grid type**, and set all other parameters as we did earlier with hexagon grid generation. Click on the **OK** button when you are done. Now, we have two algorithm blocks in the modeler's working area that use the same input as shown in the following figure:

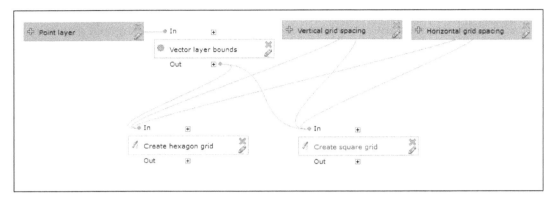

The next algorithm we should add is **Count points in polygon**. Find it in the **Algorithms** tab of the modeler window by typing its name in the filter field, and double-click on it to open the algorithm definition dialog:

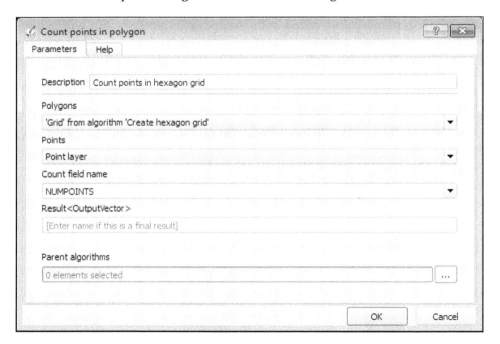

Here, we also edit the **Description** field to be able to distinguish the results of processing different grids, for example, `Count points in hexagon grid`. If you look at the available choices in the **Polygons** and **Points** comboboxes, you will see that both have an identical set of options: `Point layer`, `'Output' from algorithm 'Create hexagon grid'`, and `'Output' from algorithm 'Create square grid'`. The first is our previously defined input, and the second and third are the results produced by already added **Create grid** algorithms. As you can see, we can easily identify the results produced by algorithms, thanks to their different descriptions.

Careful readers may ask, "Why do we see both the items in comboboxes? After all, they have different geometry types and we should see only those input items that have matching geometry types." Well, this is a current limitation of Processing — the modeler does not perform any accurate checks of the input. It always shows all of the available inputs with the corresponding type (vector, raster, or geometryless table) despite defined restrictions.

So, in the **Polygons** combobox, we need to select the hexagonal grid produced by the **Create grid** algorithm (`'Output' from algorithm 'Create hexagon grid'`), and in the **Points** combobox, select the previously defined `Point layer` input. Leave the **Count field name** field unchanged. This parameter will be used only in the next step and can be safely kept hardcoded.

We again leave the **Result** field empty, as the density map produced at this stage is a temporary, intermediate result. It may contain empty cells, which should be deleted. Finally, press the **OK** button to complete the algorithm definition and add it to the modeler working area.

Add another **Count points in polygon** algorithm to create a square density map. Set its description to `Count points in square grid`, and as the **Polygon** input, select the output from `'Output' from algorithm 'Create square grid'`. All other parameters should be the same as in the previous algorithm. Also, keep the **Result** field empty.

Now, we can add the last algorithm to complete our simple model — **Extract by attribute**. With this algorithm, we will clean up our density maps from cells that do not contain any useful information. In other words, we will remove empty cells.

Find the **Extract by attribute** algorithm in the **Algorithms** tab of the modeler window by typing its name in the filter field, and double-click on it to open the algorithm definition dialog:

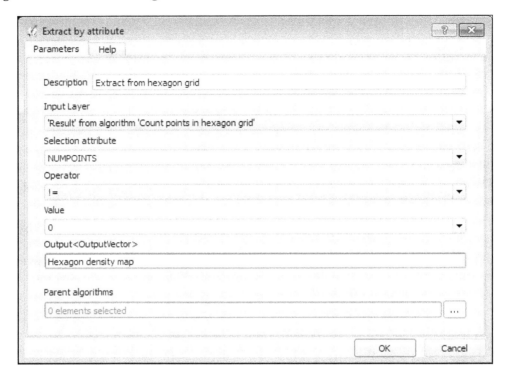

As before, we edit the **Description** field to distinguish the same algorithm blocks and their output. For the description, we choose Extract from hexagon grid. In the **Input layer** combobox, we should select the output of the **Count points in polygon** algorithm—'Result' from algorithm 'Count points in hexagon grid'. As we have hardcoded the field name in the previous step, we enter the same name in the **Selection attribute** field. Select != (not equal to) from the **Operator** combobox and enter 0 in the **Value** field.

As this algorithm will produce a result that we want to use later, it is necessary to enter a result description in the **Output** field, for example, `Hexagon density map`. Click on the **OK** button to complete the algorithm definition and add it to the modeler working area. In addition to the algorithm block, this time, an output block will be added too.

Finally, add another **Extract by attribute** algorithm to produce the final square density map. Don't forget to adjust the algorithm description, set the correct inputs, and specify the output name.

Currently, the model elements are placed in a slightly random order, and it is difficult to recognize the links between them. To improve readability and organize the model in a more structured way, we can drag all the blocks within working area. The links between them will be preserved. If necessary, you can also zoom in and out with the mouse wheel. Here is what final model may look like:

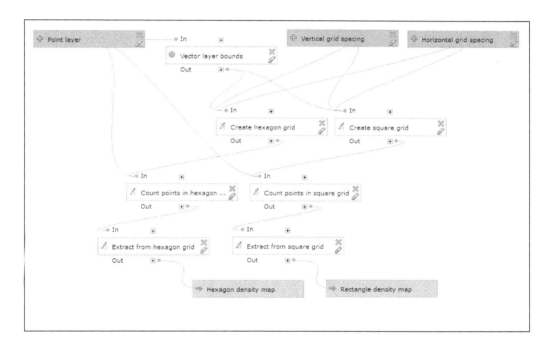

That's all! We just create our first model, and if you have layers loaded in QGIS, you can try it out by clicking on the **Run model** button on the toolbar.

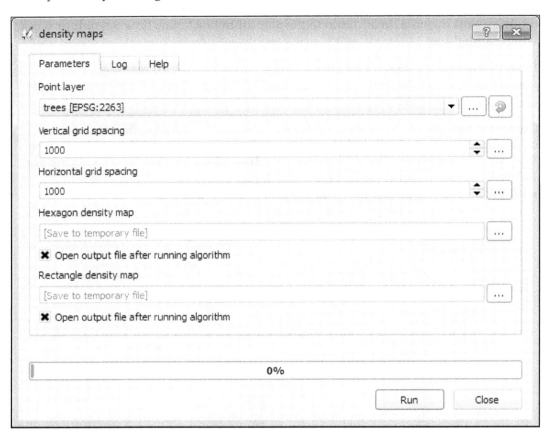

As you see, the model execution dialog has the same look and feel as all other Processing algorithms. Input vector layers can be selected from already available layers and loaded from the disk. Numerical values can be selected with spinboxes, can be accompanied by a calculator, and can maintain the restrictions we set.

But don't close the modeler right now! To fully use our model, run it from the toolbox as a batch process, and reuse it in other models, we need to save and document it first. That's what we are going to do in the next section.

Filling model metadata and saving

After creating the model, is it necessary to save it on the disk so that Processing will be able to load it and register within toolbox.

By default, models are saved in JSON format with the `.model` extension in the models subdirectory inside the processing directory in the your QGIS user folder. Under Windows, this is usually `C:\Users\login\.qgis2` (`login` here is the name of your Windows user) and under Linux, it is `~/.qgis2`. If necessary, you can always change the location of the folder using the **Processing settings** dialog.

When Processing starts, it looks for files with the `.model` extension in this directory and loads them. Loaded models appear in the toolbox, in the **Models** group. Also, they are available in modeler like any other Processing algorithm.

 Sometimes, you may get errors if you are loading third-party models. This mainly occurs because some algorithms used in a particular model are not available. For example, a certain algorithm provider is deactivated from Processing settings, or the model requires additional scripts/models, which are not available in your Processing. In such cases, read the error message carefully and examine Processing's logs to learn which algorithms are missed. Activate or install them and try to load the problematic model again.

Before saving the model, it is necessary to define its name and the group where it will be placed. This information should be entered in fields above the modeler's working area. We choose `Density` maps as the model name and `Binning` as the group name. Feel free to choose your own names if you don't like these.

When the model name and group are defined, click on the **Save** button in the toolbar and enter the name of the model. You will see a confirmation message when the model is saved.

We can close the modeler window right now, but don't rush! It is good practice to document your models, that is, describe the input, the actions performed, and the final results. Such information will be extremely useful for other users, who may want to reuse the model in their own tasks. Also, this metadata will afterwards help you recall what is this model is for.

To start editing the model metadata, click on the **Edit model help** button in the modeler dialog toolbar. A **Help editor** dialog will open, like this:

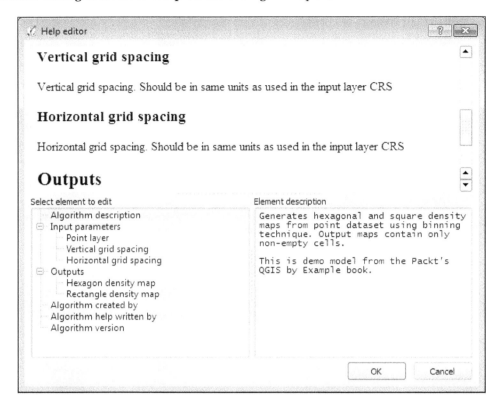

This dialog is divided into three areas. At the top, there is a preview area. Here, the current help contents are displayed so that you can see how the final result will look in real-time mode. In the bottom-left part is the elements tree, where all help sections are listed, including algorithm description as well as parameters, input, and other information. In the bottom-right part, there is an editing area. Here, we will enter a description of the corresponding element.

To edit the description of the element, select it in the elements list and enter text in the **Element description** field. To save the changes, just select another element in the elements tree.

Go through all the items in elements tree and enter their descriptions. For example, as **Algorithm description**, you can use the following text:

```
Generates hexagonal and square density maps from point dataset using
binning technique. Output maps contain only non-empty cells.
This is demo model from Packt's QGIS by Example book.
```

Describe all the other fields yourself. Try to be short and, at the same time, give as much useful information as possible. Don't explain obvious things—it's better to concentrate on important details. For example, in the description of the grid spacing inputs, it is worth mentioning that grid spacing should be defined in the same distance units as those used by layer. When you're done, click on the **OK** button to close the **Help editor** dialog.

The model metadata will be saved automatically in the same file as the model itself when you click on the **Save** button in the toolbar.

Editing models

You can also load any existing model in the modeler to edit it. You can do this, for example, to adjust some hardcoded parameters, to redefine the workflow, or just to learn how it works. There are two ways of loading an existing model:

1. Find the model in toolbox by typing its name in the search field. Right-click on the model to open the context menu and select **Edit model**.

2. Open the Processing modeler from the **Processing** menu, click on the **Open model** button from the toolbar and navigate to the model file.

To edit any input or algorithm, click on the small pencil icon in the bottom-right corner of the corresponding block. Also, you can choose **Edit** from the context menu, opened by right-clicking. Any of these actions will open a definition dialog where you can perform the necessary changes, for example, update the values or restrictions or reconnect the elements. After you have clicked on the **OK** button, the dialog will be closed and links between model blocks will be updated, if necessary.

To delete unnecessary items (input or algorithm), click on the cross button in the top-right corner of the corresponding block, or select **Remove** from the item's context menu. Note that the algorithm or input can be removed only if there are no other elements depending on it. In other words, an input should not be used by any algorithm, and algorithm outputs should not be used as inputs in other algorithms. If you try to delete a block that has items depending on it, you will see a warning message and the operation will be canceled.

Now, let's edit our model. As in almost all use cases, grid cells should have equal dimensions in both horizontal and vertical directions. It is very good to leave only one numeric parameter in the model. This simplifies the model and makes the user's life simpler, as they will need to enter less data when executing the model.

First, we need to edit one of the numeric inputs and change its name to match the input meaning. Click on the pencil icon in the bottom-right corner of the **Vertical grid spacing** block, change the **Parameter** name to `Grid spacing`, and click on the **OK** button to save your edits.

 Alternatively, you can edit the **Horizontal grid spacing** input. This replacement is completely safe.

Now, open the definition dialog of the **Create grid** algorithm and select the `Grid spacing` input in both fields: **Horizontal spacing** and **Vertical spacing**. When you're done, click on the **OK** button to save your edits and close the dialog. You will see that the connections between blocks have now changed—one numeric input (its name depends on the input you edited previously) is not connected to any algorithm. To delete this input, click on the cross in the top-right corner. The updated model may look like this:

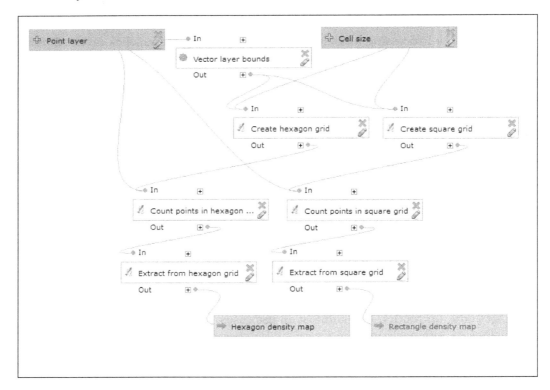

If you want, you can save the updated model as a new model. To do this, just enter a different model name and, if necessary, a group name. Then, click on the **Save as...** button and enter a name for the new model file.

It is also possible to deactivate some parts of the model—a particular algorithm or even an entire branch—without deleting the corresponding blocks. This feature is extremely useful when you don't want to get all of the output produced by the model or need to debug/test a small part of it.

To deactivate an algorithm, right-click on it and select **Deactivate** in the context menu. The corresponding algorithm block will be grayed out, and all algorithm blocks that depend on it will be automatically deactivated too. Deactivated algorithms will be skipped during model execution and will not generate any output. This has an advantage—the model's execution time is reduced. Keep in mind that the status of the algorithms (active/deactivated) is being saved in the model file, and before saving a model, make sure that you don't miss anything.

To activate an algorithm, right-click on it and select **Activate** from the context menu. Note that this activation affects only the selected algorithm. All algorithms that depend on it will remain deactivated, and you'll have to activate them one by one.

As an example, imagine that you don't need to generate square density maps. Of course, we can create a new model from an existing one by deleting unnecessary blocks and using the **Save as...** functionality. Creating a new model makes sense when you often need to generate only hexagonal density maps. But if there is a one-time requirement, it is better to simply deactivate the **Create square grid** block in our model, and all the dependent blocks will be deactivated too. So, you can execute the model, generate only hexagonal density maps, and then activate the disabled blocks again.

Sharing models

If you have created useful models that may help other users, it would be good to share them with the community so that others don't need to reinvent the wheel.

As the saved Processing model is a single file in JSON format, the easiest way to share it with others is to send it to those who are interested in it, or upload the model file to any file sharing or hosting site and provide everyone with a link to this file.

A bit more complex—but at the same time, a very convenient and user-friendly—way is to publish your model in the Processing models and scripts community repository. This repository was created in the spring of 2014, and provides a centralized way to share Processing scripts and models among QGIS users.

To put your model into this repository, you need to fork the GitHub repository (`https://github.com/qgis/QGIS-Processing`), commit your model in your fork, and issue a pull request.

 To learn more about Git, use one of the Packt's books, such as *Git: Version Control for Everyone*, and refer to the GitHub documentation at `https://help.github.com/`.

Another option is to send the model to the qgis-developer mailing list, or directly to one of Processing's developers and ask them to put it into the repository.

To get models from this repository, use the **Get models from on-line script collection** tool located in the **Tools** subgroup under the **Models** item in the Processing toolbox.

Summary

In this chapter, you learned how to use the Graphical modeler from the QGIS Processing framework to create geoprocessing models from multiple algorithms. The Modeler allows us to automate analyses and increase productivity by combining complex analyses that require several steps into a single, easy-to-use algorithm that can be reused.

We also covered additional important topics, including documenting models and sharing them with other users.

9
Automating Analysis with Processing Scripts

In the previous chapter, we introduced the Processing Graphical modeler, and you learned how to use it to automate complex geoprocessing analyses. But this is not the only way to automate your work. QGIS's Processing framework also allows you to write your own scripts in Python and use them like any other algorithm later. This is what we will discuss in this chapter.

In this chapter, we will go through the following topics:

- Python scripts in Processing
- Defining the input and output
- Implementing the algorithm
- Writing help and saving
- Sharing scripts

Python scripts in Processing

You have already learned how to create models and automate analysis tasks with the help of the graphical modeler from the QGIS Processing framework. Although the Processing modeler is user friendly and easy to use, it has some limitations:

- In the models, you can use only algorithms that are already available in Processing. Moreover, some algorithms that are available from the Processing toolbox are not available in the modeler.
- There is no support for conditional statements and loops.

So, if you need to implement something more complex and advanced, you'll need another tool. Fortunately, the modeler is not the only way to automate analyses with Processing. Processing also allows us to combine its own power with the power of the Python programming language by developing Python scripts. Such scripts can then be used like any other algorithm from the Processing toolbox or modeler, or executed as batch processes.

Before we start implementing our scripts, it is necessary to understand how to use Processing algorithms from the QGIS Python console, because this knowledge is necessary for successful use of the existing Processing algorithms within scripts.

Now, open the Python console by clicking on the button on the **Plugins** toolbar. Alternatively, you can use the *Ctrl + Alt + P* keyboard shortcut or open it by going to **Plugins | Python Console**. A new dock widget will appear at the bottom of the QGIS window, as shown in the following screenshot:

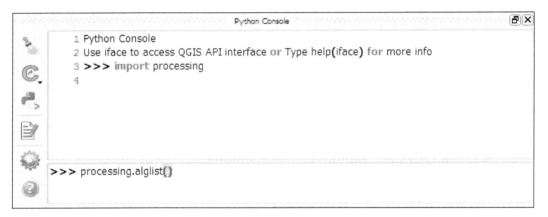

The Python console consists of two main areas. At the top, there is the output area, where executed commands and their output will be printed. Below it is the input area, where you should enter the code to be executed. Here, code is entered line by line. There is also a toolbar on the left-hand side of the Python console.

> If you want to learn more about the Python console, click on the **Help** button from its toolbar and read the built-in documentation.

To start using Processing from the Python console, we should import it with this command:

```
import processing
```

This line will load all Processing functions and make them available for us.

Listing the available algorithms

You may want to run an existing Processing algorithm from your script instead of reimplementing its functionality yourself. To do this, it is necessary to get the name of that algorithm, which is not what you see in the toolbox, but a special name — the so-called command-line name.

Every algorithm in Processing has two names: a human-readable and user-friendly name that is used in the toolbox and modeler, and another internal command-line name that has no ambiguous characters in it, such as spaces, semicolons, and so on. It also contains information about the algorithm provider. Command-line names are unique.

To list all available Processing algorithms with their command-line names, run this Python console command:

```
processing.alglist()
```

You will get a very long output that might look like this (truncated):

```
Calculator------------------------------------------
>modelertools:calculator
Raster layer bounds---------------------------------
>modelertools:rasterlayerbounds
Vector layer bounds---------------------------------
>modelertools:vectorlayerbounds
Add autoincremental field---------------------------
>qgis:addautoincrementalfield
Add field to attributes table-----------------------
>qgis:addfieldtoattributestable
Advanced Python field calculator--------------------->qgis:advancedpython
fieldcalculator
Bar plot--------------------------------------------->qgis:barplot
Basic statistics for numeric fields------------------>qgis:basicstatistic
sfornumericfields
Basic statistics for text fields--------------------->qgis:basicstatistic
sfortextfields
Clip------------------------------------------------->qgis:clip
```

To the left, you can see human-readable algorithm names that are also used in the toolbox, and to the right are the corresponding command-line names.

As the number of algorithms — even in a default QGIS installation — is really big, it may be difficult to find the command-line name of the desired algorithm. Fortunately, it is possible to reduce the output of the `alglist()` command. Just pass to it a string parameter representing a substring that should exist in the algorithm's name. For example, to display only algorithms with the word `count` in their names, execute the following code:

```
processing.alglist('count')
```

The result will be much shorter, and it will be easy to find the algorithm you are looking for:

```
Count points in polygon----------------------------
>qgis:countpointsinpolygon

Count points in polygon(weighted)-------------------->qgis:countpointsinp
olygonweighted

Count unique points in polygon---------------------->qgis:countuniquepoi
ntsinpolygon

v.qcount - Indices for quadrant counts of sites lists.---->grass:v.qcount
```

Now we know how to get the command-line name of the required algorithm. But in order to run the algorithm, we will need to know some more information.

Getting information about an algorithm

To execute an algorithm, we need not only its name but also the syntax. This includes information about the list of algorithm input and output, as well as the order in which they should be passed to the algorithm. All of this information can be obtained with the help of the `processing.alghelp()` command. This command accepts only one argument — the command-line algorithm name — and returns a list of algorithm inputs and outputs, with their types.

As an example, let's look at the `Create grid` algorithm we used in previous chapters. Its command-line name is `qgis:creategrid` (you can easily check this using the information from the preceding section), so to get information about its syntax, we should execute the next command in the QGIS Python console:

```
processing.alghelp('qgis:creategrid')
```

Here is the output of this command:

```
ALGORITHM: Create grid
  TYPE <ParameterSelection>
  EXTENT <ParameterExtent>
  HSPACING <ParameterNumber>
  VSPACING <ParameterNumber>
  OUTPUT <OutputVector>


TYPE(Grid type)
  0 - Rectangle (line)
  1 - Rectangle (polygon)
  2 - Diamond (polygon)
  3 - Hexagon (polygon)
```

From this output, we can see that the human-readable algorithm name is `Create grid` and it accepts four input fields: `TYPE` (selection from a predefined list of values), `EXTENT` (the extent), `HSPACING`, and `VSPACING` (both of these are numbers). The algorithm has produced one vector output. The most interesting part, however, is below the parameters and outputs list; it's the list of available values for the `TYPE` selection parameter. The numbers on the left are values that can be passed to the algorithm, and on the right, you can see the human-readable description of each value. For example, if you want to create a grid with the diamond cells, then it is necessary to pass a value of 2 to the `TYPE` parameter.

Now let's see how different parameter types should be passed to the algorithm:

- **Raster or vector layer and tables** (`ParameterVector`, `ParameterRaster`, and `ParameterTable`): It is possible to specify the name of the corresponding layer or table, if that layer is already loaded in QGIS. Also, you can use the path to the layer file. Finally, it is possible to pass an instance of the corresponding QGIS class, for example, `QgsVectorLayer`. If this is an optional input and you don't want to use it, just use the `None` value.

- **Selection from predefined values** (`ParameterSelection`): This should be represented by the numerical index of the corresponding value. Mapping between values and indexes is displayed as part of the `processing.alghelp()` function output, as shown earlier. There is also a separate command for listing such matches—the `processing.algoptions()` function. This command accepts only one argument—the command-line name of the algorithm—and its output is a match between the option index and value for all algorithm parameters with the selection type.

- **Multiple input** (`ParameterMultipleInput`): This should be passed as a string delimited with semicolons (`;`). Each value can be a layer name or the path to the file.

- **Field of the table** (`ParameterTableField`): This is just a string with the field name. Note that this values is case-sensitive.

- **User-defined table** (`ParameterFixedTable`): This is defined as a list of values separated by commas and enclosed in double quotes. Also, it is possible to pass a 2D list or array with the values; for example, a small 2 x 2 table can be passed as two-dimensional Python list, like this: `[[0, 1], [2, 3]]`. Keep in mind that values should start from the topmost row and go from left to right.

- **Coordinate reference system** (`ParameterCrs`): Use the EPSG code of the corresponding CRS.

- **Extent** (`ParameterExtent`): This is represented as string containing the `xmin`, `xmax`, `ymin`, and `ymax` values delimited by commas (`,`).

- **Numerical** (`ParameterNumber`), **boolean** (`ParameterBoolean`), **and string** (`ParameterString`) **parameters**: These are represented by the corresponding native Python data types: `int`, `float`, `boolean`, `str`, or `unicode`. Also, such parameters may have default values. To use them, specify `None` in the place of the corresponding parameter.

For output data, the rules are much simpler. If you want to save a layer, table, file, or HTML output in a particular place, just pass a path to the file. In the case of raster and vector output, the extension of the file will determine the output format. If the given extension is not supported by the algorithm, the output will be saved in the default format (which depends on the algorithm), and the corresponding extension will be added to the specified file path. To write the output to a temporary file, pass the `None` value.

You don't need to define any variables for numerical and string output when running an algorithm from the Python console. These will be calculated and returned automatically, without any actions from your side. See the *Executing the algorithm and loading the results* section of this chapter for more information about accessing such kinds of output.

Now that you know the algorithm syntax and how to pass parameters to it, we can execute the algorithm from the QGIS Python console.

Executing the algorithm and loading the results

To execute the algorithm from the QGIS Python console, we need to use the `processing.runalg()` method. Generally, this method is called in the following way:

```
results = processing.runalg(algorithm_name, param1,
  param2, …, paramN, output1, output2, …, outputM)
```

Here, `algorithm_name` is the command-line algorithm name, `param1...paramN` are algorithm parameters, and `output1...outputM` are algorithm outputs. Parameters and outputs should be passed in the same order as shown by the `alghelp()` method, considering information about defining input and output from the previous section.

 As we mentioned previously, you don't need to specify any variables for numeric, string, or boolean outputs.

If the algorithm reports its execution progress, a message bar with the progress indicator will be displayed in the main QGIS window during execution.

Unlike algorithm execution from the toolbox, the `runalg()` method does not load any results into QGIS. You can load them manually with the help of the QGIS API or in the following way using the helper methods provided by Processing.

On successful execution of the algorithm, the `runalg()` method returns a dictionary in which the output names (as shown by the `alghelp()` method) are the keys and their values are paths to the generated files or contain calculated values.

To load a generated raster or vector layer, pass the path to the corresponding file to the `load()` method. For example, if the result of algorithm execution is saved in the `res` variable and the output name is `OUTPUT`, then to load this layer into QGIS, execute this code:

```
processing.load(res['OUTPUT'])
```

Then, the layer will be loaded into QGIS.

To access numeric or string output in the results dictionary, just use the corresponding key names.

Let's load data from the Brooklyn tree cadastre (this is the `trees` layer in our dataset), and try to understand how the next small example works and what it does. Run these commands in the QGIS Python console one by one. If necessary, use the `processing` helper methods, such as `alglist()` and `alghelp()`, and examine the results by printing them or loading into QGIS, as mentioned previously:

```
(1) import processing

(2) resGrid = processing.runalg('qgis:creategrid', 3,
'972921.0000478327274323,1023822.9999672472476959,147696.99995
49686908722,208209.0000113248825073', 1000, 1000, None)

(3) resCount = processing.runalg('qgis:countpointsinpolygon',
resGrid['OUTPUT'], 'trees', 'NUMPOINTS', None)

(4) finalMap = processing.runalg('qgis:extractbyattribute',
resCount['OUTPUT'], 'NUMPOINTS', 1, '0', None)

(5) processing.load(finalMap['OUTPUT'])
```

If you have read the previous sections carefully, you should understand that these commands are used to generate a hexagonal density map from the point layer. In the first line, we import the `processing` module with all its algorithms. In the second line, the `Create grid` algorithm is executed, and it creates a hexagonal grid (a parameter with value equal to 3 represents hexagonal grid type) using the extent of the `trees` layer and cell size equal to `1000`. The result is saved in a temporary file, as we pass the `None` value as the last argument. In line 3, the algorithm `Count points in polygon` is executed. As a polygon layer, it uses an output of the `Create grid` algorithm (`resGrid['OUTPUT']`), and as point layer, already opened trees layer in QGIS is used. Again, the result is stored in the temporary file. Then, in line 4, the `Extract by attribute` algorithm is called to save only nonempty cells (a parameter with value 1 corresponds to the not equal to operator, `!=`). With the last line, the final result is loaded into QGIS.

Now, that you know how to get all of the necessary information about Processing algorithms and can use them from the QGIS Python console, we can dive into Processing script development.

Defining inputs and outputs

As we have already said, despite models, you can create your own Processing scripts using the Python programming language. Basically, Processing script is Python code plus some additional metadata required by Processing.

Every Processing script starts with a special block of metadata. This information is needed by Processing to register a script as an algorithm and use it from the toolbox, modeler, and so on. Each metadata entry is placed on a new line, starting with the double Python comment symbol (##), and has the following structure:

```
element_name = element_type [optional_parameters]
```

Metadata items can be divided into three groups: items that describe the script, items that describe script input, and items that describe script output.

There are only three items that describe the script:

- group: This is used to define the name of the subgroup inside the **Scripts** group in the toolbox where the script will be shown. For example, if you put the following line ##Density maps=group into the script header, it will be placed under the **Density** maps subgroup. If this item is omitted, the script will be placed under the **Scripts** subgroup, which is under **User**.

- name: This defines the script name. By default, the script name is generated from the name of the script file by dropping the extension and replacing underscores with spaces. If you don't want to use long, descriptive filenames for your scripts but still want nice names in the Processing toolbox, use the name metadata item. Its syntax is the same as that in the group item, for example, ##Hexagonal density map=name.

- nomodeler: This item is a flag. Scripts with such metadata can be used only from the toolbox. They will be not available in the modeler. Its usage is like this: ##nomodeler.

The number of metadata items that are used to describe script input is a much larger. Scripts support almost all inputs available in Processing. It is necessary to mention that the item name will also be a variable name, and it can be used in the script code. The value entered or selected by the user while executing the script will be assigned to the corresponding variable.

Also, the item name will be used as as the caption for the corresponding widget in the algorithm execution dialog. To improve appearance, underscores will be replaced by spaces. So, if you have an item with the name My_cool_parameter in the script, then its widget will have a caption as **My cool parameter**. To access the value of this parameter, we need to use the My_cool_parameter variable.

Let's look at the available input parameters:

- `raster`: This describes the input raster layer. Here is a usage example: `##Raster_layer=raster`.

- `vector`: This describes the input vector layer. Note that this item should be used if your script accepts vector layers with any geometry type—point, line or polygon. If you want to limit the supported geometry types, use one of the following items. An example of its usage is `##Vector_layer=vector`.

- `vector point`: This describes the input point vector layer. This item will accept only layers with point geometry. Note that such a limitation is applied only to layers that are already loaded into QGIS. If the user wants to specify a file from the disk, it is their responsibility to select the layer with the correct geometry type. Its usage is like this: `##Vector_layer=vector point`.

- `vector line`: This describes the input line vector layer. This item will accept only vectors with line geometry. Again, note that such a limitation is applied only to layers that are already loaded into QGIS. If the user wants to specify a file from the disk, it is their responsibility to select the layer with the correct geometry type. Its usage is like this: `##Vector_layer=vector line`.

- `vector polygon`: This describes the input polygon vector layer. This item will accept only vectors with polygon geometry. Once again, note that such a limitation is applied only to layers that are already loaded in QGIS. If the user wants to specify a file from the disk, it is their responsibility to select the layer with the correct geometry type. An example usage is as follows: `##Vector_layer=vector polygon`.

- `table`: This is used to define an input geometryless table. Its usage is like this: `##Table_to_join=table`.

- `multiple raster`: This is used to define a compound input that consists of several raster layers. An example usage is as follows: `##Layers_to_mosaic=multiple raster`.

- `multiple vector`: This is used to define compound input that consists of several vector layers. Note that this input allows us to select any vector layer, regardless of its geometry type. Its usage is like this: `##Layers_to_merge=multiple vector`.

- `selection`: This describes selection from a list of predefined values. The values specified after the item type are separated by semicolons. An example usage is as follows: `##Method=selection Nearest neighbor;Average;Cubic`.

- `boolean`: This defines a Boolean (also often called logical) input. It is necessary to specify a default value. Its usage is like this: `##Smooth_results=boolean False`.

- `extent`: This defines the input extent. Its usage is like this: `##Grid_extent=extent`.

- `file`: This is used to define an input file (in text or in any other format that cannot be recognized by Processing as raster, vector, or table). An example usage is as follows: `##Index_data=file`.

- `folder`: This describes the input directory. Its usage is like this: `##Input_directory=directory`.

- `number`: This defines a numerical (integer or floating-point) input. It is necessary to specify a default value. If the default value does not have a decimal separator, then the parameter will accept only integer values. Currently, it is not possible to define minimum and maximum limits for such parameters in scripts. An example usage is as follows: `##Width=number 1000.0`.

- `field`: This describes an attribute field in the vector layer or geometryless table. It is necessary to specify the name of the corresponding input that represents the parent layer or table. For example, if a vector layer is defined as `##Input_layer=vector`, then the field of this layer will be defined as `##Classification_field=field Input_layer`.

- `string`: This is used to define a string input. It is necessary to specify a default value. Its usage is like this: `##Field_name=string NUMPOINTS`.

- `longstring`: This defines a multiline string input. It is necessary to specify a default value. An example of its usage is as follows: `##Options=longstring my cool options`.

- `crs`: This describes the coordinate reference system. By default, EPSG:4326 is used. If you want another default CRS, specify its EPSG code. Its usage is like this: `##Assign_CRS=crs EPSG:3857`.

Input layers and tables are always passed to the script as strings containing paths to the corresponding files. To create a QGIS object (`QgsVectorLayer` or `QgsRasterLayer`) from such a string, we need to use the `processing.getObjectFromUri()` function. Multiple raster or vector inputs are also passed as strings that contain paths to individual files separated by semicolons.

Here are all the available outputs:

- `output raster`: This represents the raster layer generated by the script. Its usage is like this: `##NDVI_raster=output raster`.

- `output vector`: This represents the vector layer generated by the script. An example of its usage is as follows: `##Vector_grid=output vector`.

- output `table`: This represents a geometryless table generated by the script. This can be, for example, a CVS or DBF table. Its usage is like this: `##Nearest_points=output table`.

- output `html`: This describes the output in HTML format. Such an output is mainly used for different textual reports that may or may not include graphics. An example of its usage is as follows: `##Statistics=output html`.

- output `file`: This is used for files in formats different from HTML and all others supported by QGIS formats. For example, these can be plain text files, LiDAR data, and so on. Its usage is like this: `##Points_connection=output file`.

- output `directory`: This describes the output directory. It is mainly used for algorithms that produce many files, for example, when splitting the vector layer by attribute values. An example of its usage is as follows: `##Splitted_files=output directory`.

 Note that currently, Processing cannot load files from such output directories, even if these files are in the supported format. You need to manually open each file from the output directory.

- output `number`: This represents a numerical value generated by the algorithm. This value not saved anywhere and can only be used as the input in another algorithm. For example, one can implement script to calculate optimal cell size for the vector grid, then the numerical output from such algorithm can be used as input in the Create grid algorithm. An example of its usage is as follows: `##Maximum_value=output number`.

- output `string`: This is similar to the output number described previously. It represents a string literal generated by the algorithm. Its usage is like this: `##Select_condition=output string`.

Numerical and string outputs are also called hidden outputs (because they are not shown in QGIS and in Processing's results dialog), and are not automatically initialized with values. You should manually assign the corresponding values to them.

All other output will always be a string value with the path to the corresponding output file or directory. If user does not specify any filename, then the output will be saved to the automatically created temporary file, and the name of this file will be used as the output value.

It is worth mentioning that after successful algorithm execution, Processing will automatically load all output files in the supported formats. So, you don't need to add any `processing.load()` function calls to your script.

Implementing the algorithm

Now, when we know how to define input and output, we are ready to develop Python scripts for Processing.

There are two ways to create a script:

- Use your favorite text editor or IDE
- Use Processing's built-in code editor

Which method to use is a matter of taste and habits. In this book, we will use the built-in script editor. You can open it from the Processing toolbox. Locate and expand the **Scripts** group, expand the **Tools** subgroup, and double-click on the **Create new script** item. Alternatively, you can start typing `Create new script` in the filter field. The toolbox's content will be filtered, and you can easily locate the corresponding item.

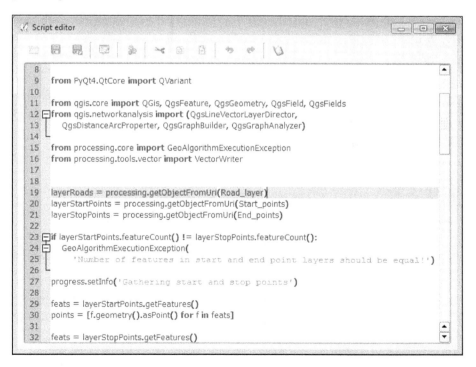

The **Script editor** looks like an ordinary text editor as shown in the preceding screenshot. There is a toolbar at the top, and all of the remaining area is used up by the editor itself.

Now we will see how to create Processing Python scripts by developing a script to find the shortest path between two points on the road network. Such an analysis is very common in many areas of application such as infrastructure planning, trip/travel planning, and so on.

QGIS has a simple built-in network analysis library that allows us to find the shortest path between two points using **Dijkstra's algorithm**, and also perform some other interesting things, such as building areas of accessibility (also known as service areas) and matching points to the nearest lines.

First, we need to determine the script input and output. As input, we need a roads layer, a start point, and an end point. As Processing does not support points as input, the simplest workaround is to pass two vectors, one for the start point and the other for the end point. This solution also allows us to define multiple start and end points and find the shortest path between each pair of start and end points. As for the output, our script obviously will have only one output — a vector layer with the shortest path (or paths) calculated.

So, the script header will look like this. If you have read the previous sections carefully, you should have no problem understanding it:

```
1 ##Network analysis=group
2 ##Shortest paths=name
3 ##Road_layer=vector line
4 ##Start_points=vector point
5 ##End_points=vector point
6 ##Shortest_paths=output vector
```

In the first two lines, we define the group where our script will be placed and the script name. Then, we define the input and output. Note that we use precise definitions for input vector layers, so that when the user runs the script in the input comboboxes, they will see only layers with the geometry type that matches the parameter definition. For example, in the Road layer combobox only line vector layers will be listed.

After the script header, a script body starts. It contains Python code that implements the desired algorithm. The script body can be divided into two parts: the import section and the code itself.

Only very simple scripts can be implemented by using Python's basic commands alone. Any more or less complex algorithm will require additional classes or libraries, for example, from the Python standard library, QGIS and Processing libraries, or even some third-party packages. To use these libraries in your code, you need to import them.

Usually, this is done on an iterative basis: you write the code, and when a new class or library is needed, add the corresponding import statement. As a rule, all import statements are placed at the very beginning of the source's file. For the sake of simplicity, in our sample script, we will provide all the necessary imports from the start. The corresponding code will look as follows:

```
(1) from PyQt4.QtCore import QVariant
(2)
(3) from qgis.core import QGis, QgsFeature, QgsGeometry, QgsField,
QgsFields
(4) from qgis.networkanalysis import (QgsLineVectorLayerDirector,
(5)     QgsDistanceArcProperter, QgsGraphBuilder,
QgsGraphAnalyzer)
(6)
(7) from processing.core import GeoAlgorithmExecutionException
(8) from processing.tools.vector import VectorWriter
```

As you can see, all imports are divided into three logical blocks with empty lines between them. In the first line, we import the `QVariant` class from the `QtCore` package of the `PyQt4` library. This class contains the definition of the universal variant data type, which will be used to declare attributes of the vector layer.

In the second block, we import various non-GUI classes from the QGIS core library (the first line of the block) and the QGIS network analysis library (the second line of the block). We need these QGIS core classes to construct our output vector features representing the shortest paths. Moreover, classes from the QGIS network analysis library provide us with all that is necessary for performing network analysis.

Finally, in the last block, we import some classes from different modules of the QGIS Processing framework. The `GeoAlgorithmExecutionException` class will be used to raise exceptions from our script, and `VectorWriter` is a helper class that allows us to easily write output vector data in any QGIS-supported format.

Now we can implement the algorithm itself. As you learned in previous sections, all our inputs—the roads layer, as well as layers with the start and end points—are passed in the form of paths to the corresponding files, so it is necessary to create layer objects from them for further usage. This done with the following code:

```
(1) layerRoads = processing.getObjectFromUri(Road_layer)
(2) layerStartPoints = processing.getObjectFromUri(Start_points)
(3) layerStopPoints = processing.getObjectFromUri(End_points)
```

We use the `getObjectFromUri()` function from the processing package. It accepts the path to the file or any other URI and returns a valid layer (raster or vector, depending on the URI) from it. Note that we specify the names of the inputs defined in the script header as arguments in the `getObjectFromUri()` function calls. As we have already mentioned in the *Defining inputs and outputs* section, the representing input value has the same name as the corresponding input.

As we use separate layers for start and end points, it is necessary to verify that both the layers have the same number of features because if the number of points is different, it will be impossible to create start-point-end-point pairs; some points will be without pairs. Here is the code for such a check:

```
(1)  if layerStartPoints.featureCount() != layerStopPoints.
featureCount():
(2)      GeoAlgorithmExecutionException(
(3)          'Number of features in start and end point layers should be
equal!')
```

We use the `featureCount()` method to get the number of features in the layers with the start and end points and compare them. If the numbers are not equal, an exception is raised and algorithm execution is aborted.

Always raise `GeoAlgorithmExecutionException` when you need to abort script execution because of any error. In this case, the user will see a standard dialog with an error message, and also the error will be stored in the log.

Often, algorithms need some time to complete, so it is good practice to inform the user about the progress of execution and provide some information about the completed steps so that the user will know that the program has not frozen. To do this, each algorithm in Processing provides a special `progress` object. With its help, you can easily display different types of messages to the user (information, debug, ordinal text, and so on), as well as show the progress of execution with the progress bar.

Our script is no exception. There are several potentially long-running tasks that should be performed during the analysis. The first is collecting the coordinates of all start and end points from the corresponding layers. These coordinates will be used later in the step of road graph generation. The corresponding code is given here:

```
(1)  progress.setInfo('Gathering start and stop points')
(2)
(3)  feats = layerStartPoints.getFeatures()
(4)  points = [f.geometry().asPoint() for f in feats]
(5)
```

```
(6) feats = layerStopPoints.getFeatures()
(7) tmp = [f.geometry().asPoint() for f in feats]
(8) points.extend(tmp)
```

In the first line, we show the user an information message with the `progress.setInfo()` command.

 There are also separate commands for displaying plain text (`setText()`), debug information (`setDebugInfo()`), console output (`setConsoleInfo()`), and other message types.

This message will be displayed in the **Log** tab of the algorithm dialog.

Then we use an iterator on the features of the layer with the start points, extract the geometry of each feature, convert it into a `QgsPoint` instance, and store it in the `points` list.

Using the same approach, we create another list, called `tmp`, with the data from the layer containing our end points. In the last line, we merge these two lists into one, so our `points` list will contain the start points and then the end points. We need to put all the points in a single list because later, all the points will have to be tied to our road network. This operation is performed for all points at once and at the same time as graph creation. As the number of points in both layers is equal, we can easily access pairs of the start and end points using very simple math. Assuming that the number of points in each layer is *N* and knowing that list indexes in Python start with *0*, we can say that the start points will have indexes from *0* to *N-1*, while the end points will have indexes will be from *N* to *2N-1*. So, if we know the index of a start point, it is easy to get the index of the corresponding end point by adding the total number of points in any input layer to the index of the start point.

As our script will produce an output vector layer, it is necessary to prepare a special object for saving features in it. Fortunately, Processing has the `VectorWriter` class, which provides us with a convenient way to save vector data in any OGR-supported format or QGIS memory layer, without writing many lines of code. Here is the code for creating such a writer object:

```
(1) fields = QgsFields()
(2) fields.append(QgsField('id', QVariant.Int, '', 10))
(3) fields.append(QgsField('startPoint', QVariant.String, '',
254))
(4) fields.append(QgsField('endPoint', QVariant.String, '', 254))
(5) fields.append(QgsField('length', QVariant.Double, '', 20, 7))
(6)
(7) writer = VectorWriter(Shortest_paths, None, fields.toList(),
(8)                  QGis.WKBLineString, layerRoads.crs())
```

In the first five lines of code here, we create an instance of the `QgsFields` container, where our attributes' definitions will be saved, and populate it. Our example script output layer will have four attributes:

- `id`: This is the integer numeric identifier of the path
- `startPoint`: These are the coordinates of the start point of the path in the (x, y) format
- `endPoint`: These are the coordinates of the end point of the path in the (x, y) format
- `length`: This is the total length of the path

In the last line, an instance of the `VectorWriter` class was created. We pass the paths to the output file defined by the user, the list of attributes that we created previously, the geometry type, and the coordinate reference system to the constructor. Note that as the path to the output file, we specify the same variable that we used in the script header to describe the output. Also, the coordinate reference system is taken from the input roads layer, so our output will be in the same CRS.

Almost all the preparation steps needed for now are done, and now we can use the QGIS network analysis library to create a graph from our roads layer. This step is required because network analysis operates with the graph and not with the vector layer. Before creating the graph, it is necessary to instantiate all the required classes and adjust the settings. The following lines of code show you how to do this:

```
(1) director = QgsLineVectorLayerDirector(layerRoads, -1, '', '',
'', 3)
(2) properter = QgsDistanceArcProperter()
(3) director.addProperter(properter)
(4) builder = QgsGraphBuilder(layerRoads.crs())
```

First, we instantiate the so-called director, which is a base class that adjusts some settings of the graph creation process. The director accepts the parameters explained here:

- The line vector layer from which a graph will be constructed.
- The index of the attribute field where the road's direction is stored. As we don't take road directions into account, we'll pass `-1` here.
- The attribute value that represents the direct road direction for one-way roads. Direct direction means that you can move on such a road only from the start point to the end point of the road. In our example, we won't use direction information, so we pass an empty string here.

- The attribute value that represents the reverse road direction for one-way roads. When a road has reverse direction, you can only move from the end point to the start point of the road. In our example, we won't use direction information, so we pass an empty string here.

- The attribute value that represents bidirectional, or two-way, roads. Bidirectional roads are the most common roads. They allow us to move in both directions: from the start to the end, and from the end to the start. In our example, we don't use direction information, so again we pass an empty string here.

- The default road direction. This argument defines how to treat roads that have no direction information in the field specified by the first argument. It can be one of these values: 1 for direct direction, 2 for reverse direction, and 3 for bidirectional roads. For the sake of simplicity, we will treat all roads as two-way roads, so we will use the value 3.

The shortest path between two points can be calculated using different criteria (in the QGIS network analysis library, these are called properters) or even their combination—length, travel time, travel cost, and so on. There is only one built-in criteria for now in the network analysis library—QgsDistanceArcProperter—which takes the road length into account. Of course, we can add our own criteria, but for the sake of simplicity in our demo script, we will use a built-in criteria. The properter is instantiated in the second line and added to the already created director in the third line.

In the fourth line, we create the so-called builder—a class that generates a graph using settings specified by the director. The only argument that we pass to the builder is the coordinate reference system we want to use. Usually, this is the same CRS as that of the input road layer.

Now that all the settings are done, we can create the graph, which will be used to find the shortest path, as follows:

```
(1) progress.setInfo('Generating road graph...')
(2) tiedPoints = director.makeGraph(builder, points)
(3) graph = builder.graph()
(4) del points
```

As graph generation is a time-consuming operation, especially on a dense road network and for a large number of start and end points, we show an information message to the user before generating the graph itself.

The most important line in this snippet is the second one, where the makeGraph() method is called. The arguments are builder, which holds all the settings for the process of graph generation, and points, which is a list of our start and end points. As points may not be located exactly on the road, it is necessary to match them to the nearest road link. This happens in the same step as graph creation, and the makeGraph() method returns a list of so-called tied points, or in other words, points that are placed exactly on the nearest road segment.

In the third line, we get the graph object itself from the builder and store it for further usage. As we don't need the original points now (all further work will be done with the tied points), we delete them in the last line to free the memory.

 More information about the QGIS network analysis library is available in *PyQGIS Developer Cookbook* at http://docs.qgis.org/testing/en/docs/pyqgis_developer_cookbook/network_analysis.html.

Now that we have our road graph and points matched to the nearest road links, we can start finding the shortest path for each start-point-end-point pair. But first, we need to perform some helper actions, as follows:

```
(1)  count = layerStartPoints.featureCount()
(2)  total = 100.0 / float(count)
(3)
(4)  ft = QgsFeature()
(5)  ft.setFields(fields)
```

The first two lines are used to prepare values for progress reporting. The progress bar displays values from 0 to 100 percent, and we need to process count pairs of points (equal to the number of features in any input point layer). Then, a single step value will be equal to 100 divided to the number of pairs.

In the last two lines, we just prepare a vector feature instance for our output routes and assign previously defined attributes to it.

Finding the shortest path is done in the loop, as follows:

```
( 1)  progress.setInfo('Finding shortest paths...')
( 2)  for i in xrange(count):
( 3)      nStart = tiedPoints[i]
( 4)      nStop = tiedPoints[count + i]
( 5)      idxStart = graph.findVertex(nStart)
( 6)      idxStop = graph.findVertex(nStop)
( 7)
```

```
( 8)      tree, cost = QgsGraphAnalyzer.dijkstra(graph, idxStart,
0)
( 9)
(10)      if tree[idxStop] == -1:
(11)          progress.setInfo('No path found from point ({:.6f},
{:.6f}) '
(12)              'to point ({:.6f}, {:.6f})'.format(
(13)                  nStart.x(), nStart.y(), nStop.x(),
nStop.y()))
(14)      else:
(15)          nodes = []
(16)          curPos = idxStop
(17)          while curPos != idxStart:
(18)              nodes.append(graph.vertex(
(19)                  graph.arc(tree[curPos]).inVertex()).point())
(20)              curPos = graph.arc(tree[curPos]).outVertex()
(21)
(22)          nodes.append(nStart)
(23)          nodes.reverse()
(24)
(25)          ft.setGeometry(QgsGeometry.fromPolyline(nodes))
(26)          ft['id'] = i
(27)          ft['startPoint'] = '({:.6f},
{:.6f})'.format(nStart.x(), nStart.y())
(28)          ft['endPoint'] = '({:.6f}, {:.6f})'.format(nStop.x(),
nStop.y())
(29)          ft['length'] = ft.geometry().length()
(30)          writer.addFeature(ft)
(31)
(32)          progress.setPercentage(int(i * total))
```

In the first line, we inform the user about next algorithm step. In lines 3 and 4, we get the next pair of the start and end points from the list of tied points. In the next two lines, indexes of these points on the road graph are obtained.

In line 8, the actual route calculation takes place. The dijkstra() method returns a tree with the shortest path from the tree root defined by the point with the idxStart index (this is our start point from the current point pair) to all other graph nodes.

From lines 10 to 23, we go through the shortest path tree and collect all the points that form a way from the end point to the start point.

After that, from lines 25 to 30, we create a line geometry from the collected points, assign it to the feature, and also set its attributes. Then the feature is passed to the writer object and stored in the output layer.

Finally, in line 32, we update the progress bar to inform the user about the algorithm's execution status.

When all the point pairs are processed, we need to carry out a cleanup and delete unused objects, such as the road graph and output writer:

```
(1) del graph
(2) del writer
```

That's all! Now you can save the script and test it by clicking on the **Run** algorithm button on the **Script editor** toolbar. The algorithm dialog will look like what is shown in the following screenshot:

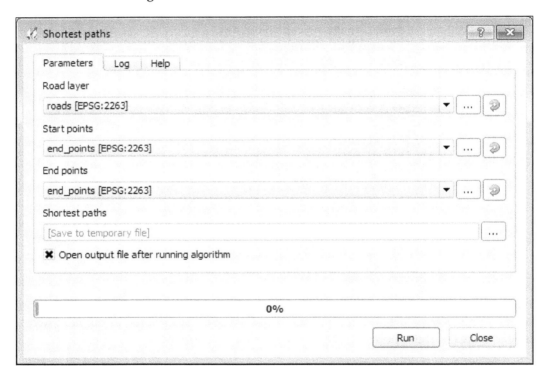

You can load the `roads`, `start_points`, and `end_points` layers from your dataset and run the algorithm with this data. Alternatively, you can use your own layer with a road network, create two point layers for start and end points, populate them with features, and execute the script. Here is what the results may look like:

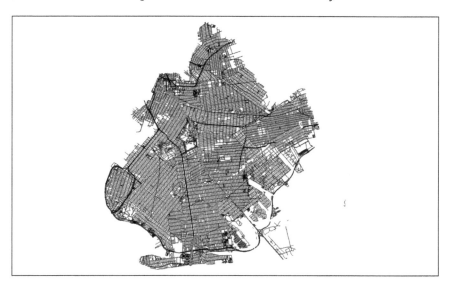

It is necessary to mention that while the QGIS network analysis library provides all the necessary tools for network analysis and can easily be extended by the user, it also has some limitations and is not suitable for work with very large and dense road networks. For such situations, it is better to use more powerful tools, such as pgRouting.

Writing help and saving

As in the case of model, it is good practice to document your scripts. A script's documentation contains its description as well as information about all inputs and outputs. This helps users to understand the purpose of the script.

To create a help file for the script, open it in the built-in Processing script editor and click on the **Edit script help** button on the editor toolbar. A **Help** editor dialog, familiar to us from the *Filling model metadata and saving* section of *Chapter 8, Automating Analysis with Processing Models*, opens.

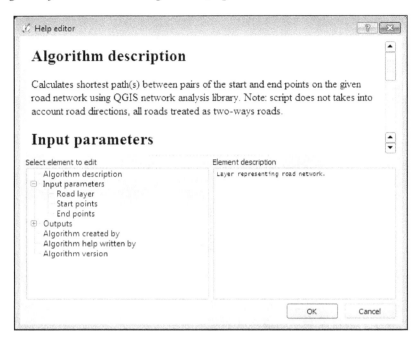

This dialog is divided into three areas. On the top, there is a preview area. Here the current help content is displayed so that you can see in real-time mode how the final result will look. In the bottom-left part is the elements tree, where all the help sections are listed, including algorithm description, parameters, input, and other information. In the bottom-right part, there is an editing area. Here, we will enter the description of the corresponding element.

To edit the description of the element, select it in the elements list and enter some text in the **Element description** field. To save the changes, just select another element in the elements tree.

Go through all the items in the elements tree and enter the descriptions. For example, for **Algorithm description**, you can use this text:

```
Calculates shortest path(s) between pairs of the start and end
points on the given road network using QGIS network analysis
library. Note: script does not takes into account road directions,
all roads treated as two-ways roads.
```

You can describe all the other fields yourself. Try to be short and, at the same time, give as much useful information as possible. Don't explain obvious things; it's better to concentrate on important details. For example, in the description of the start and end points layers, it is worth mentioning that the number of features in them should be equal. When you're done, click on the **OK** button to close the **Help editor** dialog.

To save the script help, click on the **Save** button on the **Script editor** toolbar. The script will be saved and its help will be written alongside, with the script file using the same name as used by the script with the addition of the `.help` suffix.

Sharing scripts

If you have created a useful script that may help other users, it would be good to share it with the community so that others don't need to reinvent the wheel.

The most obvious and easiest way to share a Processing Python script is just to send it to those who are interested, or upload it to any file-sharing or hosting site and make the link to this file available to everyone. It is necessary to remember that in contrast to the models, script help is stored in a separate file, not in the script itself. So, you should not forget to include the script help file when uploading or sending.

A slightly more complex way—but at the same time, very convenient and user-friendly—is to publish your script on the Processing models and scripts community repository. This repository was created in the spring of 2014, and provides a centralized way to share Processing scripts and models among QGIS users.

To put your script into the repository, you need to fork the *GitHub* repository (`https://github.com/qgis/QGIS-Processing`), commit your script and its help file in your fork in the scripts directory, and issue a pull request.

To learn more about Git, use one of Packt Publishing's books, for example, *Git: Version Control for Everyone*, by Ravishankar Somasundaram, and refer to the GitHub documentation at `https://help.github.com/`.

Another option is to send the model to the Qgis-developer mailing list, or send it directly to one of Processing's developers and ask them to put it into the repository.

To get scripts from this repository, use the **Get scripts from on-line script collection** tool, which is located in the **Tools** subgroup under the **Scripts** item in the **Processing** toolbox.

Summary

In this chapter, you learned how to develop your own Python scripts for the QGIS Processing framework and automate analysis with their help. Python scripts are an alternative to Processing's models and give us more flexibility than models. You learned how to get information about available Processing algorithms and how to call them from the Python console. Then, you got to know the main parts of a Processing script: the header with meta-information and the script body. Finally, we developed a simple script to calculate the shortest path between two points for a given road network.

In the next chapter, you will get to know another way to extent the QGIS functionality — by developing your own Python plugin.

10
Developing a Python Plugin – Select by Radius

While Processing's models and scripts are great for automating different analysis tasks, sometimes you may want to extend the QGIS functionality in another way—by developing a plugin.

In this chapter, we will go through the following topics:

- QGIS plugins
- Creating a plugin's skeleton
- Designing a plugin's GUI
- Using designer UI files in the plugin
- Implementing feature selection
- Adding translations
- Preparing the plugin for publishing

QGIS plugins

From the very beginning, QGIS was developed with the extensible and modular architecture. Originally, it allowed us to extend its functionality only with C++ plugins. But, starting from version 0.9, when Python support was added, users and developers were able to create plugins using the Python programming language as well.

Every QGIS Python plugin is just a set of the Python modules and additional files bundled into a single Python package. These packages should be placed in separate subdirectories under a special directory in the QGIS home path. Usually, this is `~/.qgis2/python/plugins`, where ~ is a user home (profile) directory. In Windows, it is `C:\Users\your_profile`, and in UNIX-like systems, it is `/home/your_profile`.

The minimal working plugin should contain two files in this directory:

- `__init__.py`: This is the package initialization file and the plugin's entry point. It should contain the `classFactory()` method.

- `metadata.txt`: This contains plugin metadata used by **Plugin Manager** and the plugins website. This metadata includes plugin name, version, description, and other information.

However, real plugins usually contain many more files in their directory: additional source files, GUI forms, corresponding sources with logic, icons and other resources, and so on. Of course, all of these files can be placed in the plugin's root directory, but to keep the source tree clean and easy to maintain, files are often organized in subdirectories. For example, Qt Designer forms are placed in the `ui/` subdirectory, corresponding sources with logic under the `gui/` subdirectory, and icons and other resources under `resources/` subdirectory and so on.

To develop QGIS plugins with Python, you will need these pieces of software:

- **QGIS**: This is meant for testing and debugging your plugin. It is better to use the same QGIS version for which the plugin is developed. If you want to develop a plugin that works on multiple QGIS versions, in all `2.x` series for example, use as old a version as possible because the newest versions may have some minor additions to the API.

- **Text editor or Python IDE**: Here, you will be writing your code. It is better to use something more advanced than the standard Notepad or any other plain text editor. Syntax highlighting, auto-indentation, and autocompletion will make your work easier and more comfortable.

- **Qt Designer**: This is used to design the user interface. For Windows, it can be installed using the `OSGeo4W` installer. The corresponding package is called **qt4-devel**. If you are a Linux user, use your package manager to find and install the Qt developer tools.

Also, to make debugging easier, we recommend that you install the **Plugin Reloader** plugin. Plugin Reloader is extremely useful because it allows you to reload your plugin after changing its code in one click, without having to restart QGIS.

In this chapter, we will develop a plugin for selecting features of the specified vector layer that are located within a given distance around reference features (already selected by the user) of another layer.

There are two ways of developing QGIS Python plugins:

1. Create a plugin template with the help of the **Plugin Builder** plugin. Then, refine this template and add the necessary functionality.

2. Develop the plugin manually by creating all the required files and code yourself.

The first approach (using Plugin Builder) is the most commonly used, and is recommended by many authors as the easiest way for novices. However, it is necessary to remember that while Plugin Builder is a great and user-friendly tool, it also hides some details, compels you to use a specific directory structure, and makes some assumptions about how your plugin will work. Also, a template generated by this plugin will contain many additional files, and these are not really necessary in all cases, for example, help file template, data for unit tests, shell scripts, and so on. Of course, all of these files can be removed or adjusted as per your needs, but it is necessary to have good knowledge to avoid deleting the necessary stuff.

In this chapter, we will create a plugin manually by adding the required files and directories step by step. This gives us full control over the plugin structure and appearance and also allows us to understand things better.

Creating the plugin's skeleton

Let's start with developing our plugin. Create a directory for the plugin somewhere in your disk. As our plugin will select features within a given radius, we call it Select by Radius and use selectradius as the name of the plugin directory.

Now, open you favorite text editor and create a file with this content:

```
[general]
name=Select by Radius
description=Select features in the given radius around another one
about=Selects features of the specified vector layer which are located
within the given radius around reference pre-selected features of the
any other layer
category=Vector
version=0.1.0
qgisMinimumVersion=2.8

author=enter_your_name_here
```

```
email=your@email

icon=icons/selectradius.svg

tags=vector,select,selection

homepage=
tracker=
repository=

experimental=True
deprecated=False
```

Save it as `metadata.txt` in the plugin directory. This is the metadata file for our plugin. As you can see, it has very simple structure, similar to INI Windows files. There is only one section called `general`, which contains all the metadata items in the `key=value` notation. Empty strings between metadata items used for logical grouping can be safely removed. The order of the metadata items does not matter as long as all the necessary items are there and their format is correct.

All metadata items can be divided into two groups: mandatory and optional. The following metadata items are mandatory and should be always presented in the `metadata.txt` file:

- **name**: This is the name of the plugin. Usually, it contains a human-readable name. Spaces and other characters such as "-" are allowed.

- **description**: This is a short description of the plugin. Usually, it is one short sentence. More detailed information should be placed in the optional "about" item.

- **version**: This is the plugin version in dotted notation, for example, `1.0.1` (if semantic versioning is used). Avoid adding words such as "version" here.

- **qgisMinimumVersion**: This defines the oldest QGIS version supported by the current version of the plugin. The value should be in dotted notation; for example, if the plugin works with QGIS version greater than 2.0, this item should have `2.0` as the value.

- **author**: This is the name of the plugin's author. Enter your name as the value.

- **email**: This is the e-mail of the plugin's author. Provide your valid e-mail address here. Note that this e-mail address is not published anywhere and is used only by the plugin repository admins if they need to contact the author.

All other metadata items are optional and can be blank. Here is the full list of optional metadata items:

- **about**: This contains more detailed information about plugin. It complements the information in the "description" metadata item.

- **category**: This is the helper metadata item. It helps users understand in which menu to look for your plugin after its installation. The supported values are `Raster`, `Vector`, `Database`, and `Web`. For example, if the plugin has to be placed under the **Vector** menu, this metadata should have the `Vector` value. If this is not set, the default `Plugins` value is used. Note that this metadata is used only as reference. You need to write the code for creating plugin actions in the correct menu by yourself.

- **qgisMaximumVersion**: This defines the last QGIS version supported by the current version of the plugin. Its value should be in dotted notation. Usually this is not used. By default, it is equal to the major number from `qgisMinimumVersion` plus 0.99. For example, if `qgisMinimumVersion` is `2.0` and `qgisMaximumVersion` is not set explicitly, it will be `2.99`. This metadata is used only in rare cases where the plugin supports a limited subset of QGIS versions or only one QGIS version.

- **icon**: This is the filename or the path to the plugin icon, if any. The path should be relative to the base plugin directory. If this is not set, the default icon will be used.

- **tags**: This is a comma-separated list of tags that describe the plugin. Try to use tags from the existing list available at the plugins website at `http://plugins.qgis.org/`.

- **changelog**: This is a list of changes in the current plugin version. It is a multiline item.

- **homepage**: This is a link to the plugin's home page, if any. We recommend filling this metadata if you plan to publish your plugin in the QGIS official plugin repository.

- **tracker**: This is a link to the `bugtracker` plugin if any. We recommend filling this metadata if you plan to publish your plugin in the QGIS official plugin repository.

- **repository**: This is a link to the plugin source code repository, if any. We recommend filling this metadata as well if you plan to publish you plugin in the QGIS official plugin repository.

- **experimental**: This is a Boolean flag used to mark the plugin as experimental. Experimental plugins may be unstable and partly nonfunctional, so they are not shown in the **Plugin Manager** unless the corresponding option is set.

- **deprecated**: This is a Boolean flag used to mark the plugin as deprecated. Deprecated plugins are not supported by authors and may not work or may work incorrectly, so they are not shown in the Plugin Manager unless the corresponding option is set.

As you can see, in our `metadata.txt` file, we have not only the mandatory items, but also some optional items to provide more information for plugin users. Note that our demo plugin has empty `homepage`, `tracker`, and `repository` metadata items. In a real plugin, especially if it will be published, these items should contain valid links to the corresponding pages so that the plugin's users can submit bug reports and patches and find the relevant documentation easily.

Also, if you look at the `icon` metadata item, you will see that it contains the relative path to the image file. So, in our plugin directory, it is necessary to create the `icons` subdirectory and put the `selectradius.svg` icon file into it. Icons can be in any raster format supported by the Qt library, but we recommend that you use PNG format for raster icons and SVG for vector icons. The icon size should be at least 24 x 24 pixels.

The next step is to create a plugin (and Python package) initialization file, `__init__.py`. This file should contain a `classFactory()` function. This function will be called when the plugin is loaded in QGIS. The function body is very short and simple:

```
(1)  def classFactory(iface):
(2)      from selectradius.selectradius_plugin import SelectRadiusPlugin
(3)      return SelectRadiusPlugin(iface)
```

The `classFactory()` function accepts a single argument called `iface` — an instance of the `QgsInterface` class that provides access to the GUI of the running QGIS copy. It returns the `SelectRadiusPlugin` object, which is a plugin instance. The code of the `SelectRadiusPlugin` class, imported from the `selectradius_plugin.py` file, is located in the plugin's root directory.

Now, let's implement the main plugin class. Create a new file called `selectradius_plugin.py` in the plugin root directory and add the following code to it:

```
( 1)  import os
( 2)
( 3)  from PyQt4.QtCore import (
( 4)      QLocale, QSettings, QFileInfo, QCoreApplication, QTranslator)
( 5)  from PyQt4.QtGui import (QMessageBox, QAction, QIcon)
( 6)
( 7)  from qgis.core import QGis
( 8)
```

```
( 9) pluginPath = os.path.dirname(__file__)
(10)
(11)
(12) class SelectRadiusPlugin:
(13)    def __init__(self, iface):
(14)        self.iface = iface
(15)
(16)        overrideLocale =
QSettings().value('locale/overrideFlag', False, bool)
(17)        if not overrideLocale:
(18)            locale = QLocale.system().name()[:2]
(19)        else:
(20)            locale = QSettings().value('locale/userLocale',
'')
(21)
(22)        qmPath =
'{}/i18n/selectradius_{}.qm'.format(pluginPath, locale)
(23)
(24)        if QFileInfo(qmPath).exists():
(25)            self.translator = QTranslator()
(26)            self.translator.load(qmPath)
(27)
QCoreApplication.installTranslator(self.translator)
(28)
(29)    def initGui(self):
(30)        self.actionRun = QAction(
(31)            self.tr('Select by Radius'),
self.iface.mainWindow())
(32)        self.actionRun.setIcon(
(33)            QIcon(os.path.join(pluginPath, 'icons',
'selectradius.svg')))
(34)        self.actionRun.setWhatsThis(
(35)            self.tr('Select features within given radius'))
(36)        self.actionRun.setObjectName('SelectRadiusRun')
(37)
(38)        self.actionAbout = QAction(self.tr('About'),
self.iface.mainWindow())
(39)        self.actionAbout.setIcon(
(40)            QIcon(os.path.join(pluginPath, 'icons',
'about.png')))
(41)        self.actionAbout.setWhatsThis(self.tr('About Select
by Radius'))
(42)        self.actionAbout.setObjectName('SelectRadiusAbout')
(43)
(44)        self.iface.addPluginToVectorMenu(
```

```
(45)            self.tr('Select by Radius'), self.actionRun)
(46)        self.iface.addPluginToVectorMenu(
(47)            self.tr('Select by Radius'), self.actionAbout)
(48)        self.iface.addVectorToolBarIcon(self.actionRun)
(49)
(50)        self.actionRun.triggered.connect(self.run)
(51)        self.actionAbout.triggered.connect(self.about)
(52)
(53)    def unload(self):
(54)        self.iface.removePluginVectorMenu(
(55)            self.tr('Select by Radius'), self.actionRun)
(56)        self.iface.removePluginVectorMenu(
(57)            self.tr('Select by Radius'), self.actionAbout)
(58)        self.iface.removeVectorToolBarIcon(self.actionRun)
(59)
(60)    def run(self):
(61)        pass
(62)
(63)    def about(self):
(64)        pass
(65)
(66)    def tr(self, text):
(67)        return QCoreApplication.translate('SelectRadius',
text)
```

In the first seven lines of this code, we import all the necessary Python packages from the Python standard library, the PyQt4 package, and the qgis.core library. Usually, these import statements are added and edited on an iterative basis during development. In other words, you write the code, and when a new class or library is needed, you add the corresponding import statement. As a rule, all import statements are placed at the very beginning of the source's file. For the sake of simplicity, in our sample plugin, we will provide all the necessary imports at the start.

In line 9, we determine the plugin's path, which will be used later to construct full paths to icons.

In line 12, a base plugin class is defined. There are several methods implemented in it. The __init__() method, also called a constructor, is used for basic initialization of the plugin instance. In line 14, we store a reference to the QGIS interface — passed as the **iface** parameter — for further usage so that we can access and use it from other methods. From lines 16 to 27, an internationalization support is activated. We check which locale is used by QGIS and try to load the corresponding translation file from the i18n subdirectory in the plugin tree. If no translation is found, the plugin will be loaded with the default locale.

We recommend that you always use English as the primary language of the plugin. Use it for all messages and captions on GUI widgets. As English is the most common and widely used language, the plugin can be used by almost all users, even without translations. If necessary, support for any other language can be added easily via the localization mechanism.

The next important, and mandatory, method that should be implemented in the plugin base class is initGui(). This method is called when the plugin is activated and loaded by QGIS. Here, we add the required GUI elements, such as menu items, toolbar buttons, and even dock widgets. We initialize all the necessary temporary folders and other stuff required by the plugin. In our demo plugin, this method starts from line 29.

From lines 30 to 36, we create a so-called action that will launch the plugin dialog. Actions are special objects that represent a command and provide a unified way to run that command from different places, such as menus and toolbars. First, we create a QAction instance and assign the Select byRadius label to it (lines 30 and 31). Note that the label text is enclosed in the self.tr() method call. This method is implemented in the last two lines of the code snippet provided, and makes the text string translatable.

Then, in lines 32 and 33, we construct an icon for our action. Using the os.path.join() call, we create a full path to the icon file located in the icons subdirectory of the plugin tree. This is the same icon that we specified in the metadata.txt file. Of course, you can use another icon—just put it into the icons subdirectory in the plugin tree. Next, in lines 34 and 35, we set the tooltip text for our action. Note that we again use self.tr() here, so this text can also be localized. Finally, in line 36, we set the objectName property of the action. This property is required by the QGIS customization framework.

From lines 38 to 42, we create another action using the same approach as the preceding one. This action will be used to show the **About** dialog with some information about our plugin. The icon for this action called about.png, and it is located in the icons subdirectory of the plugin tree.

Then, from lines 44 to 45 and 46 to 47, we add our actions to the **Select by Radius** submenu, which will be created in the QGIS **Vector** menu. In line 48, we put the button that opens the main plugin dialog into the **Vector** toolbar.

The last two lines (50 and 51) in this method are used to connect actions to handlers, which will be executed when the user presses buttons or selects menu entries. For now, both handlers—run() and about()—are empty, and we will add code to them later.

The second mandatory method that should be present in the plugin base class is `unload()`. This method is executed when the plugin is deactivated and removed from QGIS. Here, we should remove all of the plugin's GUI elements (buttons, menu entries, widgets, and so on) and perform any other cleanup actions required, such as removing temporary files. In our demo plugin, this method is defined in line 53.

As our plugin is simple enough, we just the remove menu entries added in the `initGui()` method (lines 54 to 57), as well as the toolbar button (line 58).

In lines 60 to 64, we define handlers for our actions. Currently, they do nothing.

The last method, `tr()`, as we have already mentioned, is required for internationalization support. It takes English text and returns its translated equivalent, depending on the current locale and presence of the translation file.

 We implement the `tr()` method ourselves here because the main plugin class is a pure Python class. Almost all Qt GUI classes have built-in internationalization support, and this method is already present in their code. As a result, all classes inheriting from them also will have the `tr()` method, as you will see soon.

Now, our plugin directory's structure should look like this:

```
selectradius/
├── icons
│   ├── about.png
│   └── selectradius.svg
├── __init__.py
├── metadata.txt
└── selectradius_plugin.py
```

Such a plugin skeleton can be used as the starting point for a wide range of plugins.

At this point, our plugin should be loadable by QGIS. You can easily check this — just copy the plugin directory to the QGIS plugins directory, `~/.qgis2/python/plugins`. Start QGIS and open **Plugin Manager** by going to **Plugins | Manage and install plugins...**. You should see the **Select by Radius** plugin in the **Installed** tab. After activation, the plugin will be loaded, a new button will be placed on the **Vector** toolbar, and a new entry with two items will appear in the **Vector** menu. But at this stage, these items do nothing. We need to implement their functionality.

Designing the plugin's GUI

Our plugin will have two dialogs: one is the main plugin dialog, which will be used to accept user input, and the second is the so-called About dialog, with some information about the plugin.

The Qt framework, on top of which QGIS is built, provides a special program for designing dialogs and other UI elements, such as dock widgets. It is called **Qt Designer**. Qt Designer is a user-friendly and easy-to-use visual form designer. With its help, you can create a dialog without writing code, by placing GUI widgets on the form using your mouse. The form definition in XML format is then saved in a .ui file, which is used by the plugin or application to construct a user interface.

To keep the plugin's structure clean, we will put all the .ui files in a separate subdirectory, called ui for example, in the plugin source tree.

Designing the About dialog

Start Qt Designer. In the **New form** welcome dialog, select the template called **Dialog with Buttons Bottom**, like this:

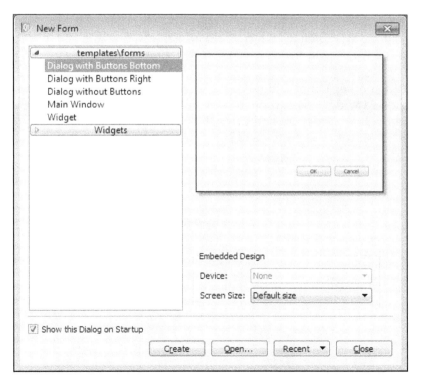

If this dialog is not opened at Designer startup, go to **File | New...** or use the *Ctrl + N* keyboard shortcut. A new empty Qt form will be created, with two buttons at the bottom.

First, we will create the About dialog. We want to display the following information in this dialog: the plugin name, icon, version number, and a short description, maybe with links to the plugin home page and/or bugtracker.

In the **Property editor** panel (usually located in the top-right corner of the Designer window), find the `windowTitle` property and change it to something meaningful, for example, `About Select by Radius`. Go back to the form and select the button box. It should be marked with blue square markers now. Return to the **Property editor**, find the `standardButtons` property and clear the checkboxes from all the variants except the **Close** button. So now, our button box has only one button.

Now, in the **Widget Box** panel (usually located to the left of the Designer window), find the **Label** widget inside the **Display Widgets** group and drag and drop it into the form. Keeping the newly added widget selected, go to **Property editor** and set the `objectName` property to `lblLogo`. This widget will be used to display the plugin icon.

 The `objectName` property of widgets will be used in our code to access the corresponding widget. So, try to assign meaningful object names for all the widgets that you plan to access from code.

Similarly, add another **Label** widget to the form and place it on the right side of the previously added widget. This time, however, don't change its `objectName` property. Instead, find the `text` property and press the button labeled **...** on the right side of the edit field. A simple text editor will be opened, like this:

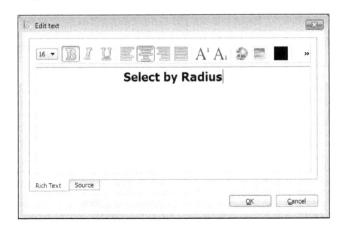

Enter the plugin name and then change the font size to a larger value, for example, 16. Make the font bold and align the text to the center. Close the editor by clicking on the **OK** button. Also modify the `alignment` property to enable horizontal alignment by center.

Add a third label, which will be used to display the plugin version, to the form and change its `objectName` property to `lblVersion`. Modify the `alignment` property to enable horizontal alignment by center. Move this label so that it will be placed under the label with the plugin name. Finally, add the **TextBrowser** widget to the form and place it under all the labels.

Qt uses a layout-based approach to manage widgets, so your form will always look consistent, regardless of the themes and fonts used. To enable layout in our form, we just have to select the form and click on the **Lay Out in a Grid** button on toolbar, or select this item from the **Form** menu.

> More information about the Qt layout system can be found in the Qt documentation at `http://doc.qt.io/qt-4.8/layout.html`. If you want to create nice-looking dialogs, this is a must-read piece of information.

Now your form should look like what is shown in the following screenshot:

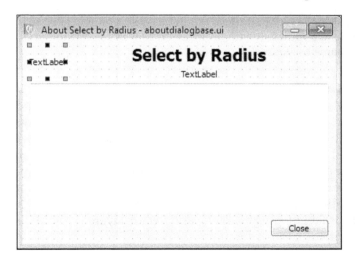

If some widgets are in incorrect places, try to move them around and adjust their sizes. When you are satisfied with the dialog's look and feel, save it as `aboutdialogbase.ui` in the `ui` subdirectory in the plugin root directory.

Designing the main plugin dialog

Close the saved **About** dialog and create a new empty form for the main plugin dialog using the same **Dialog with buttons bottom** template. Change the dialog's windowTitle property to Select by Radius and adjust the button box by changing its standardButtons property so that it contains two buttons: **OK** and **Close**.

We need to specify a target layer from which we will select features. Add to the form a **Label** widget from the **Display Widgets** section of the **Widget Box** panel, and change its text property to Select features from. Now it is necessary to provide a widget where the user can choose a layer that will be used. **Combo Box** can be a good choice here but there is an even better solution. Starting from version 2.4, QGIS provides a set of custom widgets for Qt Designer. Among these widgets, there is a special combobox called **QgsMapLayerComboBox**. It is designed to display a list of layers from the QGIS layers registry. This widget, as well as all other QGIS-related custom widgets, can be found in the **QGIS custom widgets** section of the **Widget Box** panel.

It is necessary to mention here that the QGIS custom widgets section may be not available in your system. If you cannot find it, ensure that you have installed all the QGIS-related packages (for example, the Debian package with custom widgets called libqgis-customwidgets).

Alternatively, you can use an ordinal combobox instead of **QgsMapLayerCombobox**, but in this case, you will need to implement code to populate it with layer names as well as code to retrieve a layer by its name by yourself.

Drag a **QgsMapLayerComboBox** combobox onto the form, place it on the right side of the previously added label, and change its objectName property to cmbTargetLayer. By default, **QgsMapLayerCombobox** will display raster layers, vector layers with polygonal geometry type, and plugin layers. This is not suitable for us, as we need only vector layers. To change this behavior, find the filters property and clear the checkboxes from all variants except **PointLayer**, **LineLayer**, and **PolygonLayer** (this also activates the **HasGeometry** option automatically). Now, this combobox will show only vector layers with geometry types specified previously. Raster layers, plugin layers, and vector layers without geometry will be not displayed in it.

Also, we need to specify another layer—a reference layer. So, add another **Label** widget to the form and change its text property to Around reference features from. Near this label, place the second **QgsMapLayerComboBox** combobox and change its objectName property to cmbReferenceLayer. Apply the same filters to it that we used for the previously added **QgsMapLayerCombobox** combobox.

Another input value we need is a search radius. So, put another **Label** widget on the form, under the already added widgets. Set its `text` property to `Within search radius`. On the right side of this label, put **Double Spin Box** (which can be found in the **Input Widgets** section) and change its `objectName` property to `spnRadius`. Also adjust its `minimum`, `maximum`, and `value` properties to reasonable values.

To make the plugin more useful and flexible, we provide the user with a choice of how to use the selected features: create a new selection, or alter an existing selection. The more logical way to represent available choices is a combobox. Add another label from the **Display Widgets** section to the form and change its `text` property to `Use the result to`. On the right side of this label, place **Combo Box** (which can be found in the **Input Widgets** section). Change the `objectName` property of the newly added combobox to `cmbSelectionMode`.

Put **Progress Bar** (which can be found in the **Display Widgets** section of the **Widget Box** panel) under the last label and the combobox, and change its `value` property to `zero`.

Select the dialog and apply the grid layout to it by clicking on the **Lay Out in a Grid** button in the **Designer** toolbar. Now your form should look like what is shown in this screenshot:

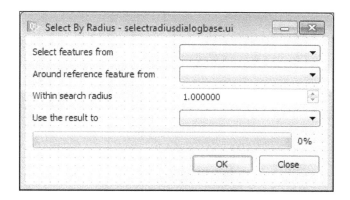

If some widgets are in incorrect places, try to move them around and adjust their sizes. Once you are satisfied with dialog's look and feel, save it in the `ui` subdirectory in the plugin root directory as `selectradiusdialogbase.ui`.

As you can see, there is nothing complex in designing a user interface with the help of Qt Designer. The most important thing here is understanding how different types of layouts work and comply with the recommendations of the **Human Interface Guidelines** (**HIG**) of the corresponding project. QGIS HIG can be found in the `CODING` document inside QGIS sources. You can view it online at `https://github.com/qgis/QGIS/blob/master/CODING#L1429`.

Using Designer UI files in the plugin

Designing the GUI with the Qt Designer is the first part of the process. Now we need to use the previously created .ui files to construct our dialogs and implement the logic required for handling user actions, such as clicking on buttons, selecting items from lists, and so on.

Adding the About dialog to the plugin

We will start from the About dialog, as it is simple. To keep the plugin directory structure clean, we will store all the sources related to the plugin GUI in the gui subdirectory inside the plugin directory.

Open your text editor and create a new file with the following content:

```
( 1) import os
( 2) import ConfigParser
( 3)
( 4) from PyQt4 import uic
( 5) from PyQt4.QtCore import QUrl
( 6) from PyQt4.QtGui import QTextDocument, QDialogButtonBox,
QPixmap
( 7)
( 8) pluginPath = os.path.split(os.path.dirname(__file__))[0]
( 9) WIDGET, BASE = uic.loadUiType(
(10)     os.path.join(pluginPath, 'ui', 'aboutdialogbase.ui'))
(11)
(12)
(13) class AboutDialog(BASE, WIDGET):
(14)     def __init__(self, parent=None):
(15)         super(AboutDialog, self).__init__(parent)
(16)         self.setupUi(self)
(17)
(18)         cfg = ConfigParser.SafeConfigParser()
(19)         cfg.read(os.path.join(pluginPath, 'metadata.txt'))
(20)         version = cfg.get('general', 'version')
(21)
(22)         self.lblLogo.setPixmap(
(23)             QPixmap(os.path.join(pluginPath, 'icons',
'selectradius.svg')))
(24)         self.lblVersion.setText(self.tr('Version: %s') %
version)
(25)
(26)         doc = QTextDocument()
(27)         doc.setHtml(self.getAboutText())
```

```
(28)            self.textBrowser.setDocument(doc)
(29)            self.textBrowser.setOpenExternalLinks(True)
(30)
(31)    def getAboutText(self):
(32)        return self.tr(
(33)            '<p>Select features of the specified vector layer
within given '
(34)            'radius around pre-selected reference features
from the another '
(35)            'vector layer.</p>'
(36)            '<p>Developed as demo plugin for the "QGIS By
Example" book by '
(37)            '<a href="https://www.packtpub.com/">Packt
Publishing</a>.</p>')
```

Save this file in the gui **subdirectory** as aboutdialog.py.

At the very beginning of the file, from lines 1 to 6, we import all the packages and classes that we will use later. Here, we use the Python standard library (the os and ConfigParser packages) as well as various PyQt classes (lines 4 to 6).

In line 8, we determine the plugin path, as we need it in order to construct full paths to the dialog's .ui file and plugin icon.

The most interesting part is lines 9 and 10, which are actually a single command split into two lines due to its length. Here, we load our previously created .ui file in the Qt Designer. The uic.loadUiType() command returns two values: our custom dialog (stored as a WIDGET variable) and its base class (stored as a BASE variable). Using uic allows us to avoid compilation of UI files and makes plugin packaging simpler.

In line 13, an implementation of the AboutDialog class, which will represent our About dialog, starts. Note that in the class definition, we use the BASE and WIDGET variables obtained previously from the uic.loadUiType() call.

The __init__() method, also called a constructor, performs a basic class initialization. In lines 15 and 16, we set up a dialog GUI. After these lines, we can access all dialog widgets using self.widgetName, where widgetName is the value of the objectName property of the corresponding widget. That's why it is important to specify meaningful and unique object names to all widgets that will be used in code.

From lines 18 to 20, we use the **ConfigParser** module from the Python standard library to read the plugin version from the metadata.txt file. Of course, we can enter the version number manually, but in this case, every time the plugin version changes, we will need to edit two files (metadata.txt and aboutdialog.py) instead of one.

Then, in lines 22 and 23, the plugin icon is constructed from the path to the icon file and then loaded into the corresponding dialog widget. In line 24, the plugin version is displayed in the lblVersion label widget.

Finally, in lines 26 to 29, we instantiate the QTextDocument object, which will be used to display the About text, and assign it to the **QTextBrowser** widget. Line 29 allows the user to open links by clicking on them.

Text for the About dialog is generated by the getAboutText() method, implemented in lines 31 to 37. The implementation is simple — we just return a string containing a short description of the plugin in HTML format. The only important thing here is the usage of self.tr() method, which allows us to show translated About text.

 For more information about Qt classes, their purpose, and their functionality, check out the Qt documentation at http://doc.qt.io/qt-4.8/index.html.

You may ask why we don't implement any methods to handle dialog execution and closing. Well, as this is a very simple dialog, we don't need to perform any special actions. We can use default handlers, which are implemented in the base QDialog class. This means that the dialog will be opened as modal, and clicking on the **Close** button will close it.

That's all! Now we need to add this dialog to the plugin main class. First, it is necessary to create an empty __init__.py file in the gui subdirectory so that Python will recognize it as a package directory.

Now, open the selectradius_plugin.py file. In the import section at the beginning of the file, add this line of code:

```
from selectradius.gui.aboutdialog import AboutDialog
```

This line makes the AboutDialog class accessible from the main plugin class. Now, find the about() method and modify it as follows:

```
(1)     def about(self):
(2)         d = AboutDialog()
(3)         d.exec_()
```

There is nothing special here. First, we instantiate AboutDialog and then execute it. Save your edits. If you want, you can update the plugin files in the QGIS plugins directory or copy the entire plugin directory here. Reload the plugin with **Plugin Reloader** and make sure that, now, when you select the **About** item from the **Select by Radius** menu, an **About** dialog is shown.

Adding the main plugin dialog

Now, let's implement the main plugin dialog. Create a new file and save it in the `gui` subdirectory of the plugin directory as `selectradiusdialog.py`. As this dialog is more complex, we split its code into small pieces, and we will examine them one by one.

As you should already know, we first import all the necessary classes:

```
(1)  import os
(2)
(3)  from PyQt4 import uic
(4)  from PyQt4.QtCore import QSettings
(5)  from PyQt4.QtGui import QDialogButtonBox, QDialog
(6)
(7)  from qgis.core import QgsGeometry, QgsFeatureRequest,
QgsSpatialIndex
(8)  from qgis.gui import QgsMessageBar
```

Besides packages from the Python standard library and PyQt classes, we also import several classes from the `qgis.core` and `qgis.gui` libraries.

Then, as in the case of `AboutDialog`, we determine the plugin path and load the dialog GUI from the Qt Designer file:

```
(1)  pluginPath = os.path.split(os.path.dirname(__file__))[0]
(2)  WIDGET, BASE = uic.loadUiType(
(3)      os.path.join(pluginPath, 'ui',
'selectradiusdialogbase.ui'))
```

The preparation is complete. Now we can define the main dialog class:

```
( 1)  class SelectRadiusDialog(BASE, WIDGET):
( 2)     def __init__(self, iface, parent=None):
( 3)         super(SelectRadiusDialog, self).__init__(parent)
( 4)         self.setupUi(self)
( 5)
( 6)         self.iface = iface
( 7)
( 8)         self.btnOk =
self.buttonBox.button(QDialogButtonBox.Ok)
( 9)         self.btnClose =
self.buttonBox.button(QDialogButtonBox.Close)
(10)
(11)         self.cmbSelectionMode.clear()
(12)         self.cmbSelectionMode.addItem(self.tr('Create new
selection'))
```

```
(13)            self.cmbSelectionMode.addItem(self.tr('Add to current
selection'))
(14)            self.cmbSelectionMode.addItem(self.tr('Remove from
current selection'))
(15)
(16)            self.loadSettings()
```

First, we initialize the dialog GUI (lines 3 to 4). In line 6, we store a reference to the QGIS interface for further usage.

In lines 8 and 9, we get references to separate buttons of the dialog's button box, as we will later need to access them as individual widgets. From lines 11 to 14, the cmbSelectionMode combobox is populated with the available selection modes. To keep the item text translatable, we wrap it into the self.tr() calls. It is worth mentioning that combobox items have zero-based numeration, so the first item will have index 0, the second item will have index 1, and so on. Populating the combobox from code allows us to easily check the order of items and determine their indices. Item indices will be used later to determine user choices.

Finally, in line 16, we restore the plugin's settings from the previous run.

Our plugin is simple enough, so there are only a few values that we want to save in the settings and restore. Every time a user opens the plugin dialog, we want to restore the previously entered search radius value and the last used selection mode from the **Use the result to** combobox:

```
( 1)    def loadSettings(self):
( 2)        settings = QSettings('PacktPub', 'SelectRadius')
( 3)
( 4)        self.spnRadius.setValue(settings.value('radius', 1,
float))
( 5)        self.cmbSelectionMode.setCurrentIndex(
( 6)            settings.value('selectionMode', 0, int))
( 7)
( 8)    def saveSettings(self):
( 9)        settings = QSettings('PacktPub', 'SelectRadius')
(10)
(11)        settings.setValue('radius', self.spnRadius.value())
(12)        settings.setValue(
(13)            'selectionMode',
self.cmbSelectionMode.currentIndex())
```

The loadSettings() method is called when we want to restore the plugin's settings. On the first run of the plugin, we have no saved settings, so the default values will be picked up. The saveSettings() method is used to save the current values from the widgets to the plugin settings.

Since we want to save the settings when the user closes the main plugin dialog by clicking on the **Close** button, and we need to start the data acquisition process when the user clicks on the **OK** button, it is necessary to replace the default handlers for these signals. The corresponding code is shown here:

```
( 1)     def reject(self):
( 2)         self.saveSettings()
( 3)         QDialog.reject(self)
( 4)
( 5)     def accept(self):
( 6)         self.saveSettings()
( 7)
( 8)         targetLayer = self.cmbTargetLayer.currentLayer()
( 9)         if targetLayer is None:
(10)             self.showMessage(
(11)               self.tr('Target layer is not set. '
(12)                       'Please specify layer and try
again,'),
(13)               QgsMessageBar.WARNING)
(14)             return
(15)
(16)         referenceLayer =
self.cmbReferenceLayer.currentLayer()
(17)         if referenceLayer is None:
(18)             self.showMessage(
(19)               self.tr('Reference layer is not set. '
(20)                       'Please specify layer and try
again.'),
(21)               QgsMessageBar.WARNING)
(22)             return
(23)
(24)         referenceFeatures = referenceLayer.selectedFeatures()
(25)         if len(referenceFeatures) == 0:
(26)             self.showMessage(
(27)               self.tr('There are no reference features
selected in the '
(28)                       'reference layer. Select at least one
feature and '
(29)                       'try again.'),
(30)               QgsMessageBar.WARNING)
(31)             return
```

The reject() method (line 1) is called when the user clicks on the **Close** button. This method was initially implemented in the base QDialog class, and we reimplement it in our subclass. The only thing we do here is save the current plugin settings (line 2). After that, we simply call the reject() method of the base class (line 3) to close our dialog.

The accept() method, defined in line 5, is called when the user clicks on the **OK** button. This method is also initially implemented in the base QDialog class and reimplemented in our subclass. When the user clicks on the **OK** button, the plugin should check whether all the required parameters are specified, find all the features from the target layer that match the defined requirements, and finally, update the selection in the target layer.

First, in line 6, we save the current plugin settings. Then, we check whether the user has selected all the necessary inputs. In line 8, we use the currentLayer() method of QgsMapLayerCombobox to obtain the currently selected target layer. This method returns the corresponding QgsMapLayer instance if a layer is selected, otherwise it returns None.

 The documentation about the QGIS API can be found at http://qgis.org/api/. Also, don't forget *PyQGIS Developer Cookbook*, which contains detailed explanations on different topics and many useful examples. You always can read the latest *PyQGIS cookbook* online at http://docs.qgis.org/testing/en/docs/pyqgis_developer_cookbook/index.html.

 Note that some samples from the latest version of the PyQGIS cookbook may not work with older QGIS versions due to API changes.

If no target layer is selected (this is checked in line 9), we show a warning message with the showMessage() method (lines 10 to 13) and return to the main plugin dialog (line 14). In other words, we don't try to perform any further actions, as we have incomplete or invalid input. The user should specify the correct input and then try again.

Using the same approach, we check whether a valid reference layer is selected (lines 16 to 22).

In line 24, we obtain a list of preselected reference features from the reference layer. If no reference features are selected in the reference layer, the length of this list will be zero. We need to catch this situation because without reference features, we cannot continue. This is what happens in lines 25 to 31; if there are no reference features selected in the reference layer, we show the corresponding message and return to the main plugin dialog.

For now, this is all of the code we need to write for the accept() method.

You may have already noticed the showMessage() method widely used in the preceding code. Here is its implementation:

```
(1)     def showMessage(self, message, level=QgsMessageBar.INFO):
(2)         self.iface.messageBar().pushMessage(
(3)             message, level, self.iface.messageTimeout())
```

This is just a wrapper that displays a message bar with the given text and importance level using the same timeout as defined in the global QGIS settings. By default, the INFO level will be used, but if necessary, we can specify any other supported level.

As you can see, we have implemented only the basic functionality of the main plugin dialog as of now.

Now, we need to add the main dialog to the plugin base class. To do this, open the selectradius_plugin.py file, if it is not opened already. In the import section at the beginning of the file, add the following line of code:

```
from selectradius.gui.selectradiusdialog import SelectRadiusDialog
```

This line makes the SelectRadiusDialog class accessible from the main plugin class. Now, find the run() method and modify it as follows:

```
(1)     def run(self):
(2)         dlg = SelectRadiusDialog(self.iface)
(3)         dlg.exec_()
```

Again, there is nothing special here. First, we instantiate SelectRadiusDialog and then open it as a modal dialog, as we did earlier for the About dialog.

Save your edits. If you want, you can update the plugin files in the QGIS plugins directory or copy the entire plugin directory here. Reload the plugin with **Plugin Reloader** and make sure that, now, when you select the **Select by Radius** item from the **Select by Radius** menu, a main plugin dialog is shown.

Implementing feature selection

Now, when dialogs are created and connected to our plugin, we can start implementing the main functionality—namely, feature selection using requirements defined by the user.

Open the `selectradiusdialog.py` file located in the `gui` subdirectory of the plugin source tree. Add the following code at the end of the `accept()` method:

```
( 1)          self.btnOk.setEnabled(False)
( 2)          self.btnClose.setEnabled(False)
( 3)
( 4)          request = QgsFeatureRequest()
( 5)          request.setFlags(
( 6)            request.flags() ^
QgsFeatureRequest.SubsetOfAttributes)
( 7)
( 8)          index =
QgsSpatialIndex(targetLayer.getFeatures(request))
( 9)
(10)          selection = []
(11)          for f in referenceFeatures:
(12)              geom = QgsGeometry(f.geometry())
(13)              bufferedGeometry =
geom.buffer(self.spnRadius.value(), 5)
(14)
(15)              intersectedIds =
index.intersects(bufferedGeometry.boundingBox())
(16)
(17)              self.progressBar.setRange(0, len(intersectedIds))
(18)
(19)              for i in intersectedIds:
(20)                  ft =
targetLayer.getFeatures(request.setFilterFid(i)).next()
(21)                  geom = ft.geometry()
(22)                  if geom.within(bufferedGeometry):
(23)                      selection.append(i)
(24)
(25)
self.progressBar.setValue(self.progressBar.value() + 1)
(26)
(27)          if self.cmbSelectionMode.currentIndex() == 1:
(28)              selection = list(
(29)
set(targetLayer.selectedFeaturesIds()).union(selection))
```

```
(30)            elif self.cmbSelectionMode.currentIndex() == 2:
(31)                selection = list(
(32)
set(targetLayer.selectedFeaturesIds()).difference(selection))
(33)                targetLayer.setSelectedFeatures(selection)
(34)
(35)                self.progressBar.reset()
(36)                self.btnOk.setEnabled(True)
(37)                self.btnClose.setEnabled(True)
(38)                self.showMessage(self.tr('Completed.'),
QgsMessageBar.SUCCESS)
```

At this moment, we have already ensured that all the necessary input fields are defined correctly (see the previous section), and now we can safely proceed with feature selection.

First, it is necessary to block the **OK** and **Close** buttons to prevent accidental clicks on them, which may interrupt running processes. We do this in lines 1 and 2.

Our plugin will need to request features from vector layers. As you know, every feature has geometry and attributes and, by default, both the geometry and attributes are returned when we query a feature from a layer. For our purposes, we need only feature geometry, so it will be better not to query attributes. This will speed up the process of feature retrieval, especially if the layer has a large attribute table or is accessed via a slow network connection.

So, in line 4, we instantiate the `QgsFeatureRequest` object, which is used to customize the process of feature retrieval. From lines 5 to 6, we alter its default behavior (fetch geometry and attributes) by resetting its `SubsetOfAttributes` flag so that only the feature geometry will be fetched.

Then, we build a spatial index for the target layer (line 8). A spatial index allows us to perform quick queries on the layer and fetch only features that intersect some region or are located close to the given coordinates. A spatial index also reduces processing time by limiting the number of features we should test against our spatial operator.

Now we are ready to search for features in the target layer that are located completely within the given radius around the reference features. But first, we need to create a list where we will store the identifiers of such features. This is done in line 10.

In line 11, we start a loop over the reference features (we already have them in the referenceFeatures list; see the previous section for details). For each reference feature, we get its geometry (line 12) and create a buffer around it using the given radius (line 13). As the buffer distance, we use a user-defined search radius set with the spnRadius spinbox. The second parameter in the buffer() call is the number of segments used to approximate curves. Bigger values will result in smoother curves and more accurate results, but will also increase processing time a bit. Feel free to change this value if you want.

Then, in line 15, with the help of the spatial index, we determine which features from the target layer may be located within the buffered geometry, which represents the current reference feature. The intersects() method returns the identifiers of all features that intersect the bounding box of the given geometry. In our case, this is the current reference geometry.

In line 17, we update the range for the progress bar. It will be used to give visual feedback about the execution of the process.

As the spatial index performs the intersection test with the bounding box, we should now test each matched feature precisely. This is done from lines 19 to 25. Let's take a closer look. In line 19, a loop starts through the matched feature indices. In the loop, we fetch a feature by its identifier (line 20). Note that we again use the previously created request to fetch only the feature geometry. Then, the feature geometry is extracted (line 21), and we check whether this geometry is located completely within the buffered reference geometry (line 22). If this condition is met, the feature identifier is added to the list (line 23).

 If you want to select features using some other criteria (for example, select features that intersect the reference geometry), replace the within() operator with the desired operator. A list of available operators can be found in the API documentation of the QgsGeometry class at http://qgis.org/api/classQgsGeometry.html.

Finally, in line 25, we update the progress bar to inform the user about the progress.

When all the reference features are processed and the identifiers of the matched features are stored in the `selection` list, we can select features in the target layer according to the requested selection mode (lines 27 to 33). First, we determine which selection mode we need to use using the conditions in line 27 and line 30. As the selection modes are added to the corresponding combobox consequentially (one by one), the first added item (**Create new selection**) will have index 0, the second item (**Add to current selection**) will have index 1, and so on. So, if the index of the currently selected item of the `cmbSelectionMode` combobox is equal to 1, then the user is asked to add features to the current selection. This means that combobox indexes match the order in which the items are added.

When the selection mode is determined, we modify the list of selected features by adding or removing indices of the already selected features (lines 28 to 29 and 31 to 32). Finally, in line 33, we select features in the target layer.

We are almost done; now it is necessary to perform some final actions. In line 35, we reset the progress bar so that it rewinds and does not show any progress. Then, we enable the **OK** and **Close** buttons so that the user can change the parameters and run the process again, or close the plugin dialog. In line 38, we inform the user that the operation is complete.

If you've made all the edits correctly, your plugin should now be fully functional and ready for testing. Just update the plugin files in the QGIS plugins directory by copying the content of your working directory, reload the plugin with **Plugin Reloader**, and test it. If there are any errors, check your code again or look at the complete plugin code shipped with this book.

Adding translations

If you followed our recommendations and used the English language for all captions in the plugin GUI, and in all strings in the code, almost all users will be able to use your plugin. Moreover, as we have made all strings translatable by enclosing them in `self.tr()` calls, it is very easy to translate our plugin into another language.

To do this, we need to prepare a so-called **project file**. This is a plain text file with a very simple structure. Create a new file and save it as `selectradius.pro` in the plugin root directory. Then, add the following content to it:

```
( 1) SOURCES = __init__.py \
( 2)          selectradius_plugin.py \
( 3)          gui/selectradiusdialog.py \
( 4)          gui/aboutdialog.py \
( 5)
( 6) FORMS = ui/selectradiusdialogbase.ui \
```

```
( 7)          ui/aboutdialogbase.ui
( 8)
( 9) TRANSLATIONS = i18n/selectradius_uk.ts \
(10)              i18n/selectradius_de.ts
```

As you can see, there is a list of all plugin source files from all subdirectories (lines 1 to 4), as well as a list of all Qt Designer UI files (lines 6 to 7). Finally, in lines 9 to 10, there is a list of translation files that will be generated (in this example, we have Ukrainian and German).

From the preceding code, you can see that we will keep translations in the i18n subdirectory, so create it if it does not exist yet.

Now open the command-line window (OSGeo shell in the case of Windows), use cd to change to the plugin directory and run this command:

pylupdate4 -verbose selectradius.pro

This will generate the .ts files required for each language specified in the project file. These files contain all the translatable strings from the plugin source code and UI forms. Using Qt Linguist, the .ts files are translated and "released." By "releasing" in this case, we mean converting a .ts file into a binary .qm file, which can be used by the Qt translation system.

It is necessary to mention that after altering strings or adding new files (sources or UI), it is a must to update the project file and regenerate the translations. Don't worry about already translated strings; they will be kept, and you will need to translate only new or modified strings.

Preparing the plugin for publishing

Once the plugin is ready and well-tested, you may want to share it with the community by uploading it to the Official QGIS Python plugins repository at https://plugins.qgis.org/.

First, it is necessary to check whether the plugin meets the following requirements:

- There is no malicious code in it
- There are no architecture-dependent binaries
- It has the correct metadata.txt file with all the required items

We have listed the most important requirements here. Other recommendations can be found at the plugins repository page and in the *Releasing your plugin* chapter of *PyQGIS Developer Cookbook*.

The next step is to prepare the plugin package. QGIS plugins are distributed in the form of ZIP archives, and each archive contains only one plugin. As plugins are extracted to the QGIS plugins directory, we must ensure that the plugin has its own folder inside the package.

Also, it is good practice to include only files that are absolutely required by the plugin in the plugin package, and omit any generated or helper files. As we load the UI files and icons dynamically, the only helper files we have are the project file and the `.ts` files (the `.qm` files should also be included because they are used by the Qt translation system). So, the content of our plugin package will look like this:

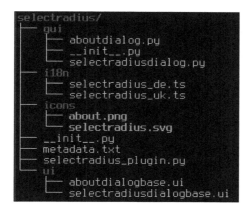

When the plugin package is created, just log in to the QGIS plugins site using your OSGeo ID. In the top menu, select **Plugins**. Then, on the left side of the page, click on the **Share a plugin** button and upload your plugin package.

 If you need to create an OSGeo ID, go to the OSGeo portal at `http://www.osgeo.org/osgeo_userid`.

That's all! Now your plugin is available for all QGIS users. Don't forget to create a bugtracker and home page with documentation about your plugin so that users can send reports about bugs and feature requests.

Summary

In this chapter, you learned how to develop QGIS plugins using the Python programming language. Python plugins allow us to extend the QGIS functionality by adding new tools or implementing new analyses or geoprocessing algorithms. You became familiar with the plugin architecture and their main components. Then we created the plugin skeleton, which can be used by many plugins. You learned how to create dialogs with the Qt Designer and use them in plugin. Finally, we developed a simple plugin for selecting features of the specified vector layer that are located within the given radius around reference preselected features of any other layer.

Index

Symbols

3D
  results, styling 173, 174
  scene general settings, working on 174, 175
__init__() method 268

A

algorithm
  implementing 247-257
atlases
  creating 97-100

B

base maps
  adding 73
  OpenLayers plugin 73
  TMS layers, adding 76
browser
  map, viewing 118
  map, working with 118-121
building footprints
  3D visualization, adjusting 177, 178
buildings layers
  combining, with DEM 157-159

C

Categorized renderer
  used, for styling vector layers 38-42
categorized vector layers
  ranking 191-195
  rasterizing 191-195
classFactory() function 266

Comma-separated values (CSV) file
  importing 17, 18
computer-aided design (CAD) 162
ConfigParser module 277
contour lines
  used, for searching distribution
         patterns 139, 140
coordinate reference system (CRS) 18, 23
currentLayer() method 282

D

database
  creating 107, 108
data, QGIS
  CSV files, importing 17-19
  GPS data, loading 20
  loading 14
  loading, from ESRI file geodatabase 16, 17
  OSM, obtaining 21, 22
  rasters, loading 16
  shapefiles, loading 14, 15
data source projection 23-26
DB Manager
  used, for deleting unused map layers 123
DEM
  combining, with buildings layers 157-159
density
  mapping, with hexagonal grid 140
density analysis 127-129
density map 128
density rasters
  ranking 191-199
Designer UI files, plugin
  About dialog, adding 276-278
  main plugin dialog, adding 279-283

Thank you for buying
QGIS By Example

About Packt Publishing

Packt, pronounced 'packed', published its first book, *Mastering phpMyAdmin for Effective MySQL Management*, in April 2004, and subsequently continued to specialize in publishing highly focused books on specific technologies and solutions.

Our books and publications share the experiences of your fellow IT professionals in adapting and customizing today's systems, applications, and frameworks. Our solution-based books give you the knowledge and power to customize the software and technologies you're using to get the job done. Packt books are more specific and less general than the IT books you have seen in the past. Our unique business model allows us to bring you more focused information, giving you more of what you need to know, and less of what you don't.

Packt is a modern yet unique publishing company that focuses on producing quality, cutting-edge books for communities of developers, administrators, and newbies alike. For more information, please visit our website at www.packtpub.com.

About Packt Open Source

In 2010, Packt launched two new brands, Packt Open Source and Packt Enterprise, in order to continue its focus on specialization. This book is part of the Packt Open Source brand, home to books published on software built around open source licenses, and offering information to anybody from advanced developers to budding web designers. The Open Source brand also runs Packt's Open Source Royalty Scheme, by which Packt gives a royalty to each open source project about whose software a book is sold.

Writing for Packt

We welcome all inquiries from people who are interested in authoring. Book proposals should be sent to author@packtpub.com. If your book idea is still at an early stage and you would like to discuss it first before writing a formal book proposal, then please contact us; one of our commissioning editors will get in touch with you.

We're not just looking for published authors; if you have strong technical skills but no writing experience, our experienced editors can help you develop a writing career, or simply get some additional reward for your expertise.

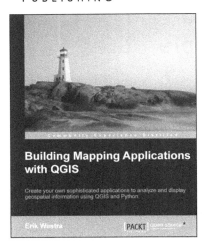

Building Mapping Applications with QGIS

ISBN: 978-1-78398-466-4   Paperback: 264 pages

Create your own sophisticated applications to analyze and display geospatial information using QGIS and Python

1. Make use of the geospatial capabilities of QGIS within your Python programs.

2. Build complete standalone mapping applications based on QGIS and Python.

3. Use QGIS as a Python geospatial development environment.

Mastering QGIS

ISBN: 978-1-78439-868-2   Paperback: 420 pages

Go beyond the basics and unleash the full power of QGIS with practical, step-by-step examples

1. Learn how to meet all your GIS needs with the leading open source GIS.

2. Master QGIS by learning about database integration, geoprocessing tools, Python scripts, advanced cartography, and custom plugins.

3. Create sophisticated analyses and maps with illustrated step-by-step examples.